By Gale Cooper, M.D.

A LABORATORY STUDY OF THE FROG

ANATOMY OF THE GUINEA PIG

INSIDE ANIMALS

ONE UNICORN

ANIMAL PEOPLE

ANIMAL PEOPLE

Gale Cooper, M.D.

with photographs by the author

HOUGHTON MIFFLIN COMPANY BOSTON 1983

Library of Congress Cataloging in Publication Data

Cooper, Gale.
Animal people.

1. Pet owners — Interviews.   2. Pet owners — Psychology.
3. Pets — Social aspects.   4. Pets — Anecdotes.   I. Title.
SF411.4.C66   1983   636.08'87'019   83-12614
ISBN 0-395-32198-0
ISBN 0-395-34838-2 (pbk.)

Printed in the United States of America

V 10 9 8 7 6 5 4 3 2 1

Photo credits: page 50, Survival Anglia, Ltd.; page 74,
Foto Baston (provided by Herta Cuneo); pages 128–29,
© Merlin D. Tuttle (Milwaukee Public Museum); page
165, Rudy Vaca.

For Lois Constantine, with love

And for Rudy Vaca, with deepest gratitude
for the help and support that
made this book possible

With special thanks to Albert Morse, Roger Caras, Kurt Benirschke, Mary Byrd, Mario Casilli, Reg Castro, Robert Dinnerman, Sharon Dinwiddie, Emilie McLeod, Suni McLeod, Michael Nelken, Beata Nelken, Ben Nelken, Miranda Nelken, Bill Patrick, Michael Pontecorvo, all the animal people, and all their animals.

The information in the interviews was received from the different animal people and has been approved by them. No claim is made for the scientific accuracy of the data included in the interviews, and the book is not intended to encourage the keeping of exotic pets.

CONTENTS

ANIMAL PEOPLE

Collecting Animal People

To you, I am nothing more than a fox like a hundred thousand other foxes. But if you tame me, then we shall need each other. To me, you will be unique in the world. To you, I shall be unique in all the world . . .

—Antoine de Saint-Exupéry, *The Little Prince*

I read *The Little Prince* as an adult, but I knew from earliest childhood the longing to experience the feeling of "to me, you will be unique in the world." And I made an unconscious decision that closeness with animals was a fulfilling way of attaining that bond.

My first memory dates to when I was about two. My nursemaid, Clarkie, was wheeling me in my big, coach-style carriage along the sidewalk outside the apartment building in the Bronx where I had lived since birth. I recall her saying to me, "If you go to sleep, when you wake up there will be the dog you want." And when I awoke, wonderful Clarkie had somehow managed to keep her promise. A shaggy black poodle-like dog was sitting beside my carriage. The appearance of that dog, which was my most longed for wish is, to this day, one of the most magical moments of my life. After a few days I was not allowed to keep it because our apartment was too small, but Clarkie gave me instead a water-filled glass globe that had a dog figurine inside. White granules floated like snow when the globe was shaken. I still have it.

No one else in my family particularly liked animals. No one owned any. New York is an environment dominated overwhelmingly by the will of man. The whole natural world exists like a zoo, with parks, trees, and even squirrels somehow all on controlled display. And I remember that when I looked out the window and rested my hands on the sill, my palms became black with soot. Nature did not press close in the Bronx. But still, what I remember most is animals. My mother taught biology and brought home a succession of frogs, guinea pigs, and pickled specimens. When I was five, we got a rabbit named Whitey, who was allowed to run loose in the house. His terminal illness several years later was my introduction to mortality. Looking back, I realize Whitey never seemed like a rabbit to me, nor did he seem like a person. He was "someone" who lived with our family.

When I was seven, my brother, my only sibling, was born. He, like Dad, kept a bemused distance from other species. But by the time of his birth a compulsion to be with animals had emerged in me, and there was never a time when I didn't have animals. Rabbits, lizards, frogs, toads, salamanders, turtles, dogs, cats, mice, rats, chinchillas, hamsters, ants, earthworms, planarians, parakeets, snakes, a monkey, parrots, a cayman, and fish (both freshwater and saltwater) were present at different times. I chose my college partly because it had an excellent horseback riding program (I had taken lessons since I was seven). I met my future husband in a pet shop, and the reason our marriage lasted as long as five years was partly attributable to our mutual love of animals. We had no children, but we did have a custody battle over the tortoises at the end.

Soon after I graduated from medical school I learned to scuba dive and went to the Mexican island of Cozumel, where Palancar, one of the world's most beautiful coral reefs, lies a few miles offshore. There I had two experiences that changed my life, ultimately leading me to the search for people who formed unique relationships with animals.

Cozumel has a tiny public aquarium made up of aquatic animals that are relatively untraumatized since they are captured in local waters and are given the same food they would normally eat. They act like animals who have no reason to fear; they are relaxed and behave normally. So when I walked along each 6-foot length of tank, an entire school of fish would follow, eyes rolling with curiosity. Then, at a display tank with young clawless lobsters, I serendipitously decided to put my hands to my lips and move my fingers like their mouth parts. To my amazement the lobsters rushed to the glass, piling one on another, urgently gesticulating with mouth parts and antennae. They seemed like lonely refugees in a foreign country who had just realized that someone spoke their language. At that moment it occurred to me that the animal world was not quite as remote as I had thought.

Then I went out on the dive boat. The first order of the day was to get lunch by traveling to beds of conch, the large aquatic mollusks whose flesh is used in a marinated dish called ceviche. A crew member would dive overboard, load a sack with the animals, climb back aboard, hammer a hole in their shells, and pull the meat out. On that day the man proceeded to throw the empty shells overboard. The captain became enraged. When I asked why, he answered, "If the conches see the shells, they will go away." And I suddenly felt that these animals, no more than a shell covering a slithering, muscular foot with some attached organs, all powered by a brain consisting of two knotlike ganglia and a nerve cord, *knew* that they were being killed. I pictured these lumpen creatures in a group on the ocean floor, perceiving the sinking shells of their murdered fellows and attempting to escape, literally at a snail's pace. Tears came to my eyes. I realized the possibility that all life forms had awareness. And I became, as a friend later said, "the woman who notices death." Because from that day onward an insect hitting the windshield of my car, an opossum dead by the side of the road, a bull dying in the sport of the bullfight, animals killed for experimentation — each in my perception died a vivid and individual death. Fur coats, leather objects, and animals used as food no longer had obscure connections to the dead individual from whom they came.

And at the same time I became aware of life. I suddenly noticed the individuality of experience of each creature. I saw birds looking around as they flew, dogs tripping or making mistakes, insects flying by me to take an interested look and then leaving. The world became filled with a myriad of individual consciousnesses, each as valid as the next.

I now live in Southern California. Behind my house is a steep hill. I like to stand at its base and watch my three all-white greyhounds — Falada, Rima, and Zanuso — playfully race across it. Their beauty is so startling that I often find it hard to believe I share my life with such exquisite beings. While I am looking I become transformed, uniting with their joyful energy. In those moments I experience a profound sense of completeness. Then I am not merely my human persona, and a part of me runs wild with the dogs, and layers of intellect, civilization, and even species boundaries peel away. And then I long to find other people who can understand how I feel.

So I began my search for people who had shared my intense involvement with animals. I wanted to feel the camaraderie of mutual passions, to meet their animals, and to learn more about their perceptions of the world. Often they called themselves animal people. I had to decide whether I was one also.

I soon realized that there were such people and that they were very special. For example, when I began my search, I heard of a woman who, as a child of ten, found an anemone in a San Francisco tidal pool. She kept it in a 10-gallon aquarium for seventy-five years, changing the seawater daily. At the end of her life she willed the creature to the world-renowned Steinhart Aquarium, where it survived for only three weeks. Perhaps it missed certain chemicals in its seawater, or perhaps it could not live without its animal person.

3

Kangaroo Mamma, Kangaroo Poppa

There are three species of giant kangaroos: the red kangaroo, the great gray kangaroo, and the wallaroo. The many smaller species of kangaroos are called wallabies. They are all marsupials, a primitive order of mammals that gestate their embryos in an abdominal pouch rather than within the womb. The baby red kangaroo, when born to its almost human-sized mother, is an inch-long pink fetus with mere stubs of hind legs. It must crawl unaided up the maternal belly into the pouch and attach itself to a waiting teat.

Violet and Murray Marcus have lived with wallaroos for many years. They have a large ranch west of Los Angeles where they breed these animals as well as a variety of cattle, sheep, and goats. They also own several acres on the bluffs overlooking Malibu Beach, with a large house in which they breed Chinese crested dogs.

VIOLET MARCUS: I was born in Spokane and lived there until I was sixteen. I had an aunt living in Hawaii, and I came to California on my way to visit her, and I met Murray. I went into a candy store, and he was running it, and he asked me how I would like to go to a lake. Being from Spokane I liked lakes, and we went out for a ride in a motorboat after he got off work at six o'clock. That is how it started.

I went to work in the candy store and he went to Compton Junior College. We ran the store as man and wife for a year and a half before my mother found out, and we had to get married. I was eighteen and he was twenty.

I had always loved animals and was an only child. When I was a little girl, my aunt gave me a couple of goslings for Christmas. I always had a dog or cat.

When Murray was ten years old, back in New York, his dad bought a spider monkey. They had a hardware store and kept the monkey in the window. It used to sleep in bed with Murray and his brother. He said, "Our folks would come home from the show and they would see this monkey lying in bed with his two big long arms around each one of our necks. They would be afraid to make a sound for fear he would bite us."

We anticipated having a family, but it turned out that I couldn't have children. We sold the candy store and got another one. After Pearl Harbor, Murray went into the ship repair unit and started to study navigation because he wanted to get a fishing boat.

We got interested in kangaroos. I wanted something I could be close to   5 and take with me. I loved deer, but you can't very well carry a deer around in a car.

I got a job taking care of a pet shop, and there was a wallaby in a stall next to a baby elephant. I fell in love with her 'cause she was so gentle. She was alone, in very small quarters, and couldn't get out to exercise. All she seemed to want was a little affection, and my heart went out to her.

Then we got a little five-month-old male wallaroo from a man who said we could buy a female in a couple of years. We named the little boy Wally. I called up an all-night talk show and I said, "If anybody out there in your audience is listening, does anybody know anything about rearing a wallaby?" The moderator never forgot that.

The hardest thing was worrying about his eating because they take so little food at the beginning, maybe a teaspoon of milk at a time, but often. When they are in their mother's pouch the teat swells in their mouth, so they don't bounce out when their mother is under way. Of course, they are just like an unborn fetus when they are born and crawl up their mother's belly into the pouch. That's the only time in their life when their forearms are longer than their hind legs.

With no children of my own, Wally filled my need to have something I could have close. When I raised him, I had an apron that I sewed together and tied around my back. When I would do my housework he was in there. I had a wicker basket to carry him in when we went to a restaurant.

We lived in Venice, which at that time was the poorest part of town. Then we bought a place in Tarzana and had a real estate office in the front with a house in the back. We accumulated some ducks, pheasants, and goats. Then Murray bought a place four blocks away that was half an acre. I kept Wally there and looked for a wife for him.

Finally we found a man who went all over the country to get animals, and he said, "Oh yes, I have a female wallaroo." When I got there, I had to buy a male llama to get the female wallaroo. That is how it really started. After the male llama was full grown we got him a wife too. The female wallaroo we had got for Wally as a wife was like me; she never had had babies. I got her when she was five months old, but she never had a family, so we got another female, who we named Wanda. Their first boy was Menahune.

Then, since Murray liked monkeys, we got a Celebes ape who was about eight months old. And after I had taken care of that elephant in the pet shop I sort of fell in love with elephants. We heard of one for sale in 1951. Murray had accumulated a little real estate at the time, so my husband bought the elephant for $2400. She was three months old, 42 inches high, and so beautiful. She had to have 2 quarts of milk all through the day and all through the night. She had been brought from Borneo. Later we were told she had to have a heated barn in winter. We weren't set up like that, so we called some people that raised animals for just traveling around in a zoo. I wish we had kept her, and we would have if we ever thought we were going to have 50 acres. I thought about it after that before I got animals, because the main thing is if you can keep it, it must be for its lifetime, 'cause every time an animal changes its home it is a trauma.

In the meantime, I was answering ads, because I was looking for a place that was a little bigger where we could get out of town. In 1969 we found 11 acres with two houses. There was a creek with about four inches of water. It cost $50,000. Then two years later the house next door came up for sale, and after that the next 23 acres with three more houses. Then we heard of a woman who was selling out her miniature traveling zoo, and we bought four horned sheep, miniature Brahma cattle, and circus wagons from her.

We went to a ranch to buy one baby deer, and we ended up getting three white-tailed deer. The boy who worked there, Danny, came to work for us and has been here for ten years.

We thought we should open to the public, but it was kind of debatable because a lot of people just want to come out and sort of go wild, and I liked quiet around the animals.

My biggest pleasure with the kangaroos was seeing them all so contented and satisfied, just lying there with their feet stretched out. Wally would be so relaxed he would just turn over and his whole stomach would be exposed. I could go in when Wally was ten years old and I would say, "Dance with Mamma," and he would put his arms around me. He stood as tall as me. He was so affectionate, yet one time I went in there to feed him and he had been fighting with his son and I didn't know it. He was mad, and he turned around and grabbed me with his hands and kicked me and took me right off my feet. He was furious because his son had just got the best of him. I had a black-and-blue spot on my stomach for a while. They really can kick.

And I loved the babies. When they are first coming out of the pouch they don't do much running around. They just bolt, go as far as they can go, and stop and turn around and come back. They will do this five or six times. If the mother makes a clucking noise, they will immediately come back and take a big dive into the pouch. She can close the pouch right up just like a drawstring. It's amazing. You can't see any opening. The baby can't poke its head out or anything. When they are relaxed, half the time the tail is hanging out and they will be down at the bottom. When they nurse, the milk gland is sort of halfway down, so they just naturally assume that position. If they become active in the sun or very hot weather, they do perspire from the elbow down to the wrist. It gets all wet, just on the inside where there is no hair.

I found a lot of satisfaction in caring for them 'cause they are so innocent. They don't have any voice; they have no vocal cords. They don't have any odor. They eat like rabbits (apples, corn, dry bread) and also what we eat — hamburger, fried chicken, and chocolate. They make pellets like rabbits. They wash after they eat.

It was because of our Wally that we bought Jack Dempsey's old house. We went over to visit Murray's mother, who lived in an apartment. The next day the landlady said, "Tell those people not to come over anymore with their dog." And Murray's mother said, "Those people don't have a dog. It's a kangaroo." And the woman said, "Well, I heard it bark." When I heard this I got mad and I said to Murray, "Let's move your mother out of there." We went up and down the beach and found Jack Dempsey's old beach house. It was two stories with five legal units. Murray always said, "Wally made me money," because we bought that property for $56,000 and sold it for $400,000.

Wally lived for fourteen years in captivity. We had thirty evolve from him and three wives in ten years. They were all our pets and they all had names. We saw smoke at eleven-thirty one morning, when we were going into town to get animal food. Murray said, "It looks pretty close — turn on the rainbirds," but after the fire got going everybody turned on their water and the force became so small that it didn't do any good. Some kid had lit a cigarette and put it in a book of matches and tossed it out of the car. It was deliberate arson. He had been doing it for weeks, but on this day there happened to be a Santa Ana wind, and it took it to our place. We were the first people to get burned out. We lost our house and six rental houses. Over 300 animals were killed — deer, llamas, the cassowarys, the scarlet macaws (one used to say "Murray, telephone"), 23 hogs (I raised one from three weeks old on a bottle and then we got him a wife and they had babies and they were all here because we never eat anything we raise), 200 wild mallard ducks, geese, blackneck swans, sixty Nubian and pygmy goats, pigeons, guinea hens, a capuchin monkey, a macaque, and the kangaroos.

The kangaroos had separate pens, 40-foot runs with houses. I had the baby I was raising in the house when the fire came. It was a firestorm, over 70 miles an hour. There was nothing they could do. The fire department had three trucks sitting out front, but they wouldn't even break out their equipment 'cause the fire was coming so fast. It was no use.

The foreman and I went down to the creekbed, which was dry because it was the end of October. We opened the gates for the animals. There was an acre that had the llamas and an acre that had the deer. But the deer ran into the fire because they were used to going into the hills, and they were asphyxiated. There was one beautiful Virginia white-tail deer that had been raised from a baby and could be called by name. She would get in the car and ride. She was a real pet. We found her a few days later. The coyotes had got to her after she was killed in the fire.

The kangaroos were all lying on the ground, the babies' heads sticking out, dead from smoke inhalation. It was heartbreaking. I wouldn't leave the place, though there was nothing to do. Murray, the hired boy, and I lived there in a camper for six weeks until the government furnished a trailer. Murray had saved the one baby roo, who was in the house. It was a concrete block house with a rock roof and it burned to the ground.

Five llamas lived through it. It was a miracle. They were in sort of a basin with the miniature Brahma cattle, and the fire seemed to go over them. The four horned sheep and the bear survived, too. An Irish wolfhound, a half-breed Dane, a police dog, and a couple of others survived by digging holes in the ground under the driveway, and the fire passed over them.

People volunteered to help, and they were here day and night picking up animals. The hogs were so heavy they had to use a bulldozer to move them. The big daddy that we raised from a baby was about 1500 pounds. He burned for a couple of days because we had no water; the pipes had melted and fallen over. For months after I could smell it. It's funny how the flesh-burning smell got in your lungs and didn't leave you.

The Red Cross came and brought us sandwiches the first day, and the World Pet Society came the following day and brought us toothbrushes, toothpaste, a little hand mirror, a wind-up alarm clock — things we never even thought of. The telephone company strung an emergency telephone line because they knew Murray had a heart condition. It was like losing a family, especially when you don't have any family, and when you get older like we are, in our sixties, this was our family.

The year after the fire, I heard about a zoo that was closing in Arizona, and we bought two males and one female wallaroo. We had two baby llamas born last year and a baby bull Brahma, and we expect the pygmy goats to have babies. The roo that Murray saved from the fire was a relative of Wally's, the very first one I raised. After it was all over, she gave birth to Tina and she is pregnant again.

Tina jumped out of her pouch when she was five months old. I wanted to raise her, and we have had her with us every day and night for the past six months. She sleeps in between both of us. Sometimes she stretches out. When she is hungry she gets up, and if I don't hear her she will come and pull my hair with her mouth. Believe me, it wakes you up in a hurry. I have her bottle in bed. I get up at about a quarter to five and Murray stays in bed maybe another half-hour. So I get her some oats and bring her some fresh milk.

When roos get ten to twelve months old they want to play and be chased. Actually, the males aren't fully developed until eighteen months, but the females are. I think Tina would be interested already, but I wouldn't put her in with a male before she is a year and a half old. She doesn't need any companionship. She is with us all the time, but when you put them back with all the rest, their needs are satisfied with their own and they lose any desire for companionship with people.

As long as I live I will raise kangaroos. I don't have any reasons not to. They aren't any problem. They don't have any obnoxious habits. They don't make any noise. Who is going to complain about it? We have 50 acres and we put them in the middle and own both sides. Murray loves it here and says this is his Catskills because it looks like the Catskills without the green. But in March and April it's green. After the rain the grass gets a foot high and the yuccas bloom.

Every morning Murray and the hired boy go between 6:00 and 10:30 A.M. to get discarded fruit and lettuce trimmings from the markets. We still have to buy alfalfa and rabbit pellets. At night, instead of going in and watching TV we sit out in the swing and watch the airplanes go by because we just happen to be right under the flight pattern. I guess some people might think this is all crazy, but it is our life.

A $10,000 Koi Named Inazuma

It is difficult to find a fish person. Though millions of people have fish, most fish keepers view their aquariums and the inhabitants as some form of decoration. And while they often have an awareness of color patterns, scientific names, and even behaviors, what is missing is the ability of the person to feel any emotional attachment to the individual fish. At one point I located an elderly man who lived with his mother and a small fish called a bichir. He had been with his mother all his life and with the fish for twenty-five years. He implied that he liked neither one too well.

Then I realized that someone participating in the Japanese tradition of raising large colored carp, or koi, might cherish a fish. And I found that person in Sus Yamanaka, the director of the Zen Nippon Airinkai, a koi fanciers' society in Los Angeles. He is sixty-six, and he and his wife, Yuriko, have been married for more than forty years and have two sons. Their home, set back from the road, is distinguished from the others in the neat residential neighborhood only by the Japanese lantern at the base of the driveway. The soul of the house is in the fenced-in back yard with its koi pond and huge water-filtration system. The concrete pond is irregular in form, about 18 feet by 25 feet, and holds 12,000 gallons. The natural rock border and the dark color inside make it seem deeper than its actual 5½ feet. Eight powerful underwater jets create constant currents.

At the side of the pool are two square concrete tanks, the biological filters, covering a 6-by-15-foot area. The first is a settling tank, from which Mr. Yamanaka removes debris every morning. The second processes the smaller impurities. A total of 3000 gallons passes through the system every three hours.

High over the pond is a lath roof, filtering the sunlight and patterning the water. In the depths of the pond, gold, white, red, and black forms of almost fifty koi, some over 30 inches long, flashed past. And for the purpose of the photograph, Mr. Yamanaka graciously netted and lifted his most prized fish, Inazuma, out of the water and held him in his arms.

SUS YAMANAKA: I guess the history of koi in Japan goes back to the Buddhist temples. In the old days only the rich people could afford to have fish, and they were just river carp with not much color on them, maybe brown or all red. Only a hundred and sixty to seventy years ago they started breeding multicolored koi, called Nishiki, meaning "brocaded," like fabrics of different colors. But the origins of Nishiki koi go back four or five hundred years. I understand that they came from China, then migrated to Iran and through Europe.

I think the reason they picked carp originally is that they did have colors and they were real graceful swimmers. Also, they are supposed to have a lot of strength. They are known for swimming upstream against the current. And every now and then they will try to jump up against a waterfall. So for the people they are a symbol of grace and strength.

In Japan the oldest koi live to over a hundred or hundred and fifty years in the monasteries. One famous red and white named Hanako, or "flower," died about three years ago at two hundred and twenty-seven years. She came down through a family and lived in a big artesian lake.

My father was a berry farmer in California. He and my mother were born in Japan. I was born in the United States. They went back in 1927 and took me and my seven brothers and sisters to see their old house in Japan. At the corner of the yard was an artesian well that my dad had dug into a small mud pond. He had caught brownish river carp for it. My mother liked it too. I imagine that that is what got me started.

I remember that when I was still a kid, in California, my father and his friends used to go together on Sundays to a river that had black carp. I would tag along. They never fished with a pole. They would go in there with bathing trunks, hand-catch them, and put them in a gunny sack. In Japan they do that a lot. I never could figure out how they could catch them, and I was too young for them to teach me.

As long as I remember, I had always wanted a little fish pond. In 1967 I was working for a nursery specializing in indoor plants; I was married and had two boys. With the help of my older son, I built my first pond in the corner of the yard. It was only about 1500 gallons, a 5-by-8 concrete rectangle, and I only had about eight koi. After two or three years I built the second, which was about 5500 gallons. I had that for about ten years. Now I have just completed a pond that is about 12,000 gallons, with a filter and settling tanks of about another 3000 gallons. It is four feet deep and there are about forty-five to fifty koi. I use jets of water to simulate strong currents and make the fish grow better.

The first fish I got was a red and white. That is the most popular among fish breeders or fish fanciers. In the beginning I used to buy anything I liked. As time goes on you learn to appreciate better fish, and I gave away all the early fish to my friends. I hardly have any of the original ones. I remember twelve years ago seeing a Shusui that was about 15 inches long. That kind has large blue scales on its back. I asked the dealer how much and he said, "Two hundred dollars." I bought it; that was big money in those days.

But now I have one that cost me $10,000. In 1973, when I got it, it was only about 15 inches long and around three years old. It was already a famous fish in Japan, where it was the cover of a book on famous koi. Its owner had died and the missus couldn't take care of it. A fellow I know in Japan bought the whole pond and sold the fish to a dealer in America. For a year I kept asking to buy it, but it was sick with fin and tail rot. But I knew I could take care of it by keeping it in a separate tank with air bubbles and sulfide. The day I took him home I felt good. I watched and watched him. The pond is next to my bedroom, and every time I heard a splash I would stick my head out to make sure he wasn't jumping out of the pond. (Until they get acclimated they can jump out.) He is the most expensive fish I ever bought. At first the missus didn't like that I spent the money, but I told her I didn't think I would spend that much anymore.

The pattern of his color is real nice. When you judge a fish it is 50 percent conformation, 25 percent color, and 25 percent pattern. His color pattern is special, like a red lightning bolt or zigzag over the top of his back. I named him Inazuma, which means "the lightning bolt," and even made a license plate for my Avanti that says INAZUMA. In the first show I took him to I got the Grand Champion with him. I had had him about a year and a half and he was about 19 inches. The funny part was, there were two judges who came from Japan, and they looked at him funny, like they remembered him from somewhere. I kept quiet, but finally it came out.

These days, a fish that would take a Grand Champion in Japan could sell for $60,000 to $80,000 and are famous in the country. I remember that in a show, one of the judges had several famous fish. He had a $50,000 one shipped from Japan so he could exhibit it in America. But it was packed wrong and died. I was one of the people who tried to revive it. We gave him oxygen in the water and worked on him from 4:00 P.M., right after we got him out of Customs, to two-thirty in the morning. We think it must have

13

been several hours before he was put on the plane in Japan and the water in his plastic bag overheated. Often they put dry ice on the bottom of the box, but that time they didn't. It was a thirty-three-inch fish, a really nice fish.

I got started in showing fish about ten years ago and have been a judge in about fifteen shows. I am presently regional director of the Zen Nippon Airinkai, which can be translated as the All-Japan Koi Lovers' Society. It is one of the largest koi clubs in the United States. In judging, the fish are divided into six categories according to size. If there are very big fish, a seventh category is added. The category of very large koi is rare in the United States.

There are also thirteen different colors. In the judging, each tank has a group separated by size and color. Some common examples of colors are the *kohaku*, or red and white. This is the most popular with fanciers. There is the *taisho sanke*, or black with white with red spots, and the *bekko*, which is red and white with black on the back and on the pectoral fins. In a show the judge looks for conformation, color, pattern, and movement.

Though the fish are seldom handled, this is done sometimes for demonstration in shows. You hold one hand under the chin and one on the belly. It takes skill. They wiggle and are real slimy. You have to hold loosely and go with the wiggle or they will fall. If you hold too tightly they will squirm all the more. Once I watched a master fish handler from Japan. "Your hands have to be real limber," he said, "and just follow the fish."

Someday I might want to get into breeding, but for now I just buy, but never a fish under 7 inches. In Japan, they say, "He who sells a fish under 7 inches is stupid, and he who buys a fish under 7 inches is also stupid." That is because they can change so much as they grow. There are problems with breeding. A lot of times after breeding they will lose their color for unknown reasons. Also, they take quite a beating. The female's belly is big with eggs, and the male swims right in and hits her with his nose and knocks the eggs out. Sometimes she is all cut up. Also, the male can miss the female and ram his nose against the cement and kill himself. In Japan they float nylon netting in the water, the eggs stick, and the male fertilizes them. The spawning season is from April till June, and I try to separate my males and females then.

I am not sure about how intelligent the koi are, but some will come to you more easily than others at feeding time. The gold ones are most tame. I did have one about three years ago who used to come right to the edge of the pond when he wanted food. He was platinum with a bit of yellow tinge. I got him when he was small and had him about five years. I became attached to him. It was a funny thing, he was always right there. I think he could tell by the vibration of my footsteps. But someone must have come into the yard and killed him. He had a big cut on his head and was dead when I got him out of the water.

My oldest son is a dentist and has three kids already. The youngest works for the school board. The middle one wants to build a pond. The grandchildren like to watch the fish. Two years ago I had two small fish that I entered in competition under my grandchildren's names. They came in second, and I gave each of them a trophy.

A lot of times I come home from work and walk right up to the pond and stand there for hours just watching and feeding the fish. It is very relaxing. My wife must think I'm crazy. But I have talked to several other people who keep koi, and they tell me the same thing. They all spend hours just watching them and you forget everything.

A "Hands-on" Trainer of "Nonhuman" Beings

Ken Decroo, a Hollywood animal trainer and stunt man, says one of his mottoes is: "Dramatize life and live it with courage." At thirty-six, he has been living up to his ideal. He has a Ph.D. in linguistics and was an early member of the revolutionary Washoe Project, which involved teaching primates sign language. He now teaches sign language communication at Crafton Hills College; at Cal State San Bernardino he teaches animal behavior and exotic animal training in the Psychology Department and a course on the use of animals in the classroom in the Education Department. Most recently, he started the Wild Animal Training Center, to teach students his philosophy of animal training and caretaking. Further, he is designing a B.S. program in animal behavior at Cal State San Bernardino.

After his marriage broke up, he moved onto the compound at the Wild Animal Training Center. There his trailer is perched on bluffs overlooking the Santa Ana River, which he calls "the last wild river in Southern California." His 38 acres are in the middle of the Hidden Valley Wildlife Area, between Riverside and Norco. Living with him are a squirrel monkey and a Celebes macaque.

The animals at the center provide students with hands-on experiences. There are six chimpanzees, a young jaguar, a white-handed gibbon, six dromedary camels, raccoons, opossums, otters, a breeding pair of red foxes, a wallaby, skunks, European brown ferrets, lop-eared rabbits, rats, two lions, a black leopard, ostriches, a zebra, and a capuchin monkey. There are a variety of reptiles; among them are pythons, red-tailed boas, monitor lizards, and desert tortoises. Birds of prey, parrots, and macaws are also present.

He says that if he ever gets discouraged, all he has to do is look out the window of his trailer to feel that he is "living in a magic land."

KEN DECROO: My grandfather came to Miraloma from Oklahoma one year before the Dust Bowl hit. He was one of the lucky people who sold his farm instead of walking off it. His definition of an "Okie" was anyone who came   15

to California from Oklahoma after he did. My mother was born on the new farm; her brother and sister were brought across from Oklahoma. They had what was called a family farm. They worked it themselves and subsisted off it. I can remember my grandma and grandpa milking their four cows every night. He and I were inseparable. I used to do chores on the farm with him. I remember him walking around in blue coveralls with a hammer hanging out one side and a pliers out the other. He made me a small tool kit, a miniature of his own. He was tall, like I am now. He didn't have any teeth except on Sunday, when he put on his best khakis, pressed and shiny, and his black polished boots. He would get in his 1940 Chevy truck and let me ride on the running board. We would drive to the Adobe Inn, where he got two quarts of beer. Then he would drive back home and sit on the porch and drink his beer all afternoon, waiting for the boys from the Miraloma Improvement Association, which was organized for drinking beer and pitching horseshoes.

One of my first experiences with animals was when I was about six and I found a spider. I had it clutched in my hand, and my grandfather saw two little black legs coming out betwen my fingers. He said, "Boy, what you got in your hand?" I said, "I got me a bug." And he said, "Well, just let me see that a minute," and he took the pliers from his coveralls and took hold of both legs very gently and opened my hand. It was a black widow. My grandfather took it and put it over the fence, and said, "Since it didn't bite you, I can't kill it."

He had an incredible way with animals and greeted them as individuals. When he walked by a group of horses, he would say, "Hi, Jeffy. How you doing, Buck?" And all the horses would come over to him. Us kids would be trying to catch a horse for two hours, and he could tell it to come over and it would walk right up to him. Later it used to be said that I had the same way with animals, that I could get them to do things for me when other people had to use force. My grandfather taught me how to read animals, without calling it that or anything. We had some wonderful times.

He had a running argument with his neighbors, and I know that some of the argument was based on whether animals could feel or not. Now in those days castration practices were very cruel and my grandfather wouldn't allow that. I also remember what we called butcher day, when kids always got out of going to school. You didn't really need to help, but you watched the whole morbid process. My grandfather would never let me stay. He also had a graveyard for all his pets.

My grandmother bred parakeets and cockatiels. My grandfather and I made a zoo. He helped to design and build the cage. It started when a barn owl got injured and I kept it. Then I got snakes, scorpions, foxes, skunks, lizards, a raccoon. I even had an octopus once that someone brought me from Mexico.

17

I remember when my grandfather died. I was very young, and I had stayed home from school because I had pinkeye. My sister and I were fighting in the backyard when my grandmother came running out, screaming, "Your grandfather is dead." Both my parents' house and my grandfather's house were right on the border of our farms, maybe a quarter of a mile apart. I ran up there as fast as I could into the bedroom and he was definitely dead. Then I ran 5 miles to a nurse we knew. We didn't have a phone.

My father always called me a wild boy and said there was no fear in me. He had had a rough life. His father died when he was nine, and from then on he basically raised his brothers and sister. He grew up in Pennsylvania. He was very physical, a smart man without much formal education. My mother and he met during the war when he came to California on his way to the South Pacific. After the war they married. I was born in Pennsylvania, but they came back to the farm when I was an infant. My father was basically a city boy and a very independent man. He never farmed and became a contractor.

My dad was very physical with me. I remember an argument when I was in high school. I was going up the steps and he was yelling at me, saying, "Why did you come home so late?" I said, "None of your business." He had taught me how to throw a left jab, bob, weave, and throw a hard right hook, so that is what I tried to do. Anyway, he is not a real big man, but he had been a Golden Gloves middleweight champion in Pennsylvania. So I bobbed, weaved, and went to throw my right, and the lights went out for me. He was sad the next day, and my mom was all over him. She was furious.

Rodeo was very important to me. When I was very young we would catch calves and ride them. We also had this great big old cow that didn't milk anymore. She would really buck when I put a belt on her like a regular bull rider would. Mom and Dad would come out and watch. Mom would worry, but sometimes I think Dad didn't because he hadn't grown up on farms enough to know I really could get hurt.

I wanted to be Larry Mahem, the rodeo star. He typified all of my values. He was kind to animals. He was kind to women. He was kind to kids. You could walk up to him and he would say, "What's your name, little pard?" "My name is Kenny, sir." "Be sure and eat your cereal and mind your parents." If Larry Mahem said you should do those things, that was manly.

It was more intense hero worship than baseball provides. It had to do with the animals, I think. If you saw Larry Mahem ride a horse, you could see he was communicating with the horse. There were animals that were as much a hero as Larry Mahem. Tornado, for one. He was a bull, a big bull, and tough. Freckles Brown was one of the only men ever to ride Tornado.

18   We kids had no fear. We built tree houses in eucalyptus trees that were

60 feet high, planted as windbreakers in the twenties. The most elaborate I built was a three-tiered one about 30 feet off the ground. You climbed up with ropes and boards nailed to the trunk. We had no fear of heights. My sister, who was three years younger, was the toughest of all. I used to do horrible, wicked things to her. We had constant rock fights with big rocks. It's a wonder we didn't kill each other.

I think I do well with chimpanzees today because I grew up in a "chimp colony." We were big farmboys and settled everything by fighting. As a kid I probably broke my arm four or five times. There was lots and lots of getting sewed up by the local nurse.

We played "suicide bikes" with the idea that one guy got his heavy American Flyer on the hill, and the other guy rode at a right angle on the cross street on the bottom of the hill. The idea was to try to run him over at the cross street. We would dive from the highest rock into the reservoir. I remember that the scariest thing to do was to walk across a round water pipe that ran fifty feet off the ground across a ravine, which had only about two feet of river wash. There was also a field with not-at-all-friendly Brahma bulls. The idea was to run across the open fields with this thundering herd behind you. A lot of this I never told my mother, but my dad would just sit there and relish it. His basic principle was, "You should never let fear make you stupid." By that he meant you shouldn't do some stupid thing just to overcome something you were afraid of.

Once my friend Roger shot me. I wanted to get in the tree house and he wouldn't let me. He was shooting at my feet to make me stop. They took me to the nurse, and her husband yelled at Roger, "You shouldn't have rifles before you are twelve." We had interesting morals. The other kids didn't hate Roger for shooting me, but just for shooting me when I was unarmed.

When I was twenty I went into the army, to Vietnam. I don't talk very much at all about the army. I would just say that it further confirmed that I have new tolerances of fear. Less scares me now than it did then. After being close to death so often, facing a tiger is not that frightening to me. I developed a present-moment philosophy, and I think that most people that come out alive with all their parts working have that same basic feeling: everything to experience exists at this moment. When I am under stress or something hits me, I go through the same routine every time. I touch my body and I say, "It is all working and I am alive. It can't be terribly serious." Or when somebody tries to drop authority on me I immediately will think, "You might be able to take my job, but you can't kill me."

When I came back, my parents had moved to Riverside and sold the farm. My dad said, "A boy ought to get an education nowadays." My grandfather couldn't read. But Dad said, "Don't ever let education get in the way 19 of you being a man." He meant a man should be strong both emotionally and physically, and that he should always value and respect physical work.

I went through college at U.C. Riverside and took a dual degree in anthropology and geology. My dad had become superintendent for building trades and was doing work at a school for the deaf. I met a deaf girl there and we became really tight for about a year and a half. In that time I learned sign language 'cause that was all she knew.

I then got my Ph.D. in linguistics, writing a dissertation on the language acquisition of deaf infants. I had met my wife, Louise, who was doing research with macaque monkeys. When my dissertation came out I got a letter of invitation from Trixi Gardner, who was working with her husband, teaching sign language to chimps. I was aware of their research but was fairly skeptical. I went up for an interview and stayed for two years. I gave a talk on a Thursday evening. The next morning they told me to "feel at home and walk around the ranch." I was walking across a pasture and came to a road. All of a sudden in the trees above me I heard this "ooh-ooh-ooh." I looked up and this young chimpanzee came running down the tree. I looked around kind of nervous; I had never worked with chimps before. I figured that he had broke loose. He ran up to me. I was staring at him and he stared at me. I'll never forget the eye contact. It was intellectual. It was different. He signed, "Who you?" In total shock, I stared at him for a second. I was of course fluent in sign language. Then all of a sudden I realized that a member of another species had just asked me who I was! I said, "My name Ken." He kept on looking around as if he was doing something naughty, and he met my eyes again and went, "You me, hide and seek." I was shocked again, but I did the only thing you could do as a linguist. I said, "Who's It?" And he goes, "You It, me hide." And off he went. At that point I knew I couldn't leave. I thought it was really the most exciting thing that would ever happen to me.

After a time I became less interested in the linguistics and more in the actual training and handling of the chimpanzee. I began to see training as a very sophisticated form of communication. At that time you would read in the literature that animals were like "soft machines," without any awareness. Every day my work was telling me just the opposite. Once I got in an argument with one of the chimps. She went up a tree and she wouldn't look at me. It's like putting your hands over your ears when you are communicating by sign language. Finally, I got angry, and I decided, "I'm going to jerk her right out of this tree." Of course she weighed 130 pounds and was probably seven times stronger than me. But I'd lost my temper, and I jerked the lead as hard as I could. She turned around and looked at me and started pulling the lead up. She lifted me five feet off the ground and had me dangling there. Of course I couldn't let go of the lead or she'd have been in Reno terrorizing the casinos. Finally, as she dropped me she pointed down and made a classic chimp laugh.

Now chimps can blow up and get very aggressive. She had bitten me before this by lying — she promised to be good and I walked within reach and she nailed me. It's definitely premeditation. One day she was throwing things at me and trying to bite me. So I picked up an empty plastic garbage can and threw it. Then I picked up some rocks and threw them. Then I walked over and grabbed hold of the jeep and looked directly at her. She went into a full-face grimace and started crying and signing "Good, good." She thought if I could throw all that other stuff, maybe I could throw the car. She didn't exactly know what I could do. From that point she would set me up. She would point to stuff and have me lift it. She thought she was being very subtle, like, "Let's play." Well, my awareness was shifted by all these interchanges. Even after I was separated from her for a year, she immediately remembered my name sign. A very unique rapport develops between you and those animals.

While I was working with the primates, I was contacted by Marble Arch Productions and asked to work as a technical adviser on a movie about chimpanzees. I'll never forget the first day I saw a very good cat trainer, walking a huge Siberian tiger. It was close to 600 pounds. I watched this power moving and the trainer moving with it. It was so silent, and so different from a relationship with a chimp. Chimps are noisy animals, and you find yourself being noisy even in sign language. And you are always arguing over something. It's just the way primates are. But the cat was so intense. You could almost see the electricity coming off the lead chain back and forth.

I began to work with Tammy Maple and made an agreement to teach her chimp training if she would teach me cats. There is a prejudice in the business about women trainers not being strong enough, but as I see it, when an animal weighs 600 pounds, it's not going to matter much at all whether you weigh 200 pounds or 100 pounds.

Another friend started teaching me how to do stunts with animals, like running and letting a cat knock you down and act as if he is eating you. At that point I was in absolute heaven because it was a combination of everything I wanted out of life. It paid well, it was exciting and different all the time, and it was a nice combination of the physical and mental. You have very little time to get the behaviors. For example, in the movie *The Wild and the Free* we had eleven chimpanzees, a leopard, a 20-foot reticulated python, and ferrets. Obviously we couldn't teach sign language to the chimps in five weeks, so we had to teach behaviors that looked like a flow of actual signs. Yet one smart male figured out the language. One day he was in the cage and I was about to take him out and he looked at me and went "drink" and pointed to a glass of water.

I began to see tremendous difference in trainers. Some were very heavy-handed. There were horrible things. Hotshots or cattle prods were put on a pole to poke an animal and make it jump rather than to train it to go from point A to point B. The large animal factories were the worst because they would have hundreds and hundreds of animals. To them they were a commodity to make money. I knew a trainer once who had a chimp that would run to him with an open-mouthed play face. This trainer was shooting him with a slingshot because he thought it was aggressive behavior. Now when a chimp wants to kill you he has a dead poker face with tight lips. But the guy wouldn't believe me. You can imagine what it was doing to the chimp.

Basically, I teach my studio animals how to learn, not just a lot of unrelated behaviors. I teach them their job. Once they have that down I can get a large piece of behavior at once. My assumption is that animals have awareness.

To me, the way to teach behavior is by being "hands on" with an animal and to learn the motivations behind its behavior. The feeling I get is that the animals I work with are "nonhuman beings" — in other words, there is no ranking involved. You have to adjust for the different worlds of experience. There is the act of training and the art of training, and the art involves interspecies communication with a two-way channel.

With an exotic animal, where you haven't had ten thousand years of breeding for temperament, most of your energy in the beginning goes into forming the rapport, the relationship. Then you can go into what I call nuts and bolts, the use of conditioning or whatever. My greatest reward of training is that window of perception I get into their world. To work the great apes, animals of such high intellectual level, you have to be able to feel them and move in a rhythm. Their experience is very different from a person's. What might on the surface feel like inattentiveness might be the animal listening to a sound or feeling a breeze that's exciting him or her. In a Clint Eastwood movie I worked with Clyde, an orangutan, and they are pretty hard to read. In fact, he once really set me up in a scene. We were in a kitchen fight scene, and I was a cowboy walking alongside Clyde. I saw him sort of shifting his eyes, and then he gave me a hug. Well, he had been looking around the kitchen and saw this pot of spaghetti. He never looked at it again. Then the cameras were rolling, the fighters went through the door, and as Clyde and I started through at the last possible point, he grabbed a huge handful of spaghetti and jammed it into his mouth. So I learned to look at what he wasn't looking at.

In big cat training you are working with pre-wired behavior, with the world of instinct and reaction. It's so intense. It puts me close to the edge and gives me a surge of energy. I feel I am doing, feeling, and sharing
22   something few people know about or have ever felt. Yet it is also a kind of

meditation, a special kind of complete concentration, and the energy flows back and forth between me and the cat. The danger is certainly there. I saw a movie of a man killed by a leopard in a zoo — the entire mauling took three frames of the camera. When you watch, it looks like the leopard just ran up over his head, but he bit twelve times, two of them fatal.

I don't think anything gets your reactions as fast as a cat. A friend showed me a videotape of when I got in trouble with a tiger. I was running and he was supposed to jump and knock me down. But somehow I felt he hit a little too hard, a little too deliberate. I knew he was going to try to get me. I hit the ground on my back and the next second I was standing. On the video it looks like I had a spring in my back. Every muscle in me just tightened and I sprang to a standing position.

For a cat, it's movement that will trigger off behavior, and he will react before cognition comes in. We humans, like chimps, will do the opposite by first deciding, then reacting. When I walk a big cat on a lead I become very aware of a different world, with enormous amounts of movement and life that I couldn't see without him. Plus the whole world is gone. Suddenly you are forced to anticipate and react as the cat does or else you will be doing "cat skiing" as he drags you along behind him. In the culture of training, you never let go of the lead.

Two people always work cats. There's the chain person, who's in charge, and the cane person, who has a large hickory cane, which is used if the cat takes you down in a real life-and-death situation. I've only had to hit a cat once. He bit me, and my back-up man threw me a cane and I hit him on the nose. The best protection in an absolute emergency is a fire extinguisher.

Lions are extremely social animals. Since they can kill with one bite, they have developed strong inhibitions. The tiger, a nonsocial animal, doesn't have nearly that degree of restraint. On the other hand, you are more likely to get bitten by a lion because they use their mouths to communicate. It gets to be a very complex thing.

The ultimate in "interactional synchrony," or getting into the right rhythm, is working several cats together. You have to read off the cats you can see to know what the others are doing. For example, when I am working with two lions, and the eyes of the one in front become large and dilated and its movments become elongated and slow, I know I am being stalked by the one behind and I can react accordingly. So you are constantly looking for cues. You can tell if a cat is bluffing even if he is roaring a full scream right at you with his ears down. If his hair's not up, if he's not pyloerected, he's bluffing. From the moment I walk up to a cage I am looking for anything unusual. Any blood, how's the stool, is the cage wire torn, is he pacing, does he jump up to try to get high ground, does he come up and greet me?

When you're dealing with reptiles and amphibians it is not so much 23 training as taming and managing the environment. If you cool a snake down, he'll stay in one place. If you warm him up, his activity level will rise. If you want a snake to crawl from point A to point B, you put him in the shade and put a hot light to one side for him to crawl to. When I worked on the movie *Cannery Row,* which called for thousands of frogs, we could make them move away by shining a hot lamp and make them jump happily by sprinkling water on them. Not a single one died in the film, and we released them after the movie.

I have just started a school to turn out a whole new breed of trainers. It is called Wild Animal Training Center, with a full-time staff of five trainers. There we teach the natural behavior of animals and behavior in captivity. It is one of the few places in the world where people can get hands-on practice ranging from big cats to primates to reptiles. We teach the whole process of animal care and maintenance and the often-neglected art of transporting animals well. We are also working on movies with the strict limit that animals be depicted with dignity and accuracy. At present I also use my show biz work when I teach my physical anthropology class. It's fun to be able to let the students get hands-on with a chimp.

My latest project done with private money is training a Capuchin monkey to give a quadriplegic man the dexterity and mobility he lacks. It's called Project Harpo because, as a baby, the monkey looked like Harpo Marx. It is the most challenging training I have ever done because I want Harpo to conceptualize his role as a helper. He feeds the quadriplegic, combs his hair, retrieves his mouth stick, opens drawers, fetches objects, turns the light on or off, and runs the stereo. He can work the man's Apple computer, which controls many household devices. If the project succeeds, I will give the man Harpo for life in return for his help. And unlike seeing-eye dogs, which live for about ten years, Capuchin monkeys can live for thirty years.

There is a whole spectrum of what is perceivable in the world, and only a part of it is perceivable by *Homo sapiens.* In forming a relationship with animals, our perception extends beyond ourselves. We can take advantage of another animal's perception when it's communicated to us. We have a responsibility to animals, not dominance, but stewardship. If you put a person "hands on" with a lion, it's unlikely that the person will ever go on a safari to shoot one. I like that idea.

Wild Bird Rescue and the Birthright of Freedom

Birds represent about 20 percent of all vertebrates and about 1 percent of all animals. Taffy Beauvais, with the help of volunteers, is doing her part to protect the bird population of the San Fernando Valley of California, where she lives. In 1974 she founded the Wild Bird Care and Rehabilitation Fund and turned her home into a rescue haven. She keeps a bird list, but unlike birdwatchers, who note sightings, her record represents her wild bird rescues. She has returned to the wild an incredible variety of birds, ranging from murres to meadowlarks to birds of prey.

Taffy, her husband, Pierre, and her son and daughter live in a residential neighborhood of the San Fernando Valley. Much of the house and a good deal of the 50-by-150-foot backyard are devoted to the rehabilitation of birds. The bathroom, service porch, and even some of the bedrooms are used for birds that need extra care or warmth. Spring to early summer is the busiest season for bird rescue. The house is literally filled with berry baskets stuffed with tissue paper or shoeboxes with dry sphagnum moss — surrogate nests for multitudes of sick or abandoned baby birds. Conventional bird cages are used for small species.

The backyard has five 5-by-5-foot aviaries, mounted on wheels for mobility. Young hawks and owls live in wooden cages, since they tend to damage their wings on metal bars. Adult raptors perch on specially welded wrought-iron perches with block perches, since many like to stand on a flat surface.

The birds receive individual care. The babies are hand-fed with modified syringes. In the course of a year Taffy feeds her birds sixty thousand mealworms, hundreds of jars of baby food, both liquid and powdered vitamins, and dog kibble soaked in water.

TAFFY BEAUVAIS: The birds here at the Rescue station ultimately have to work for a living. If I tame them and they assume that people bring their food and put it in little dishes for them, they won't survive. They would never know what it is to be what they are: free creatures.

Our society has been pet-oriented for years. Consequently, any animal that enters someone's life is seen as a pet. And baby birds are so beguiling. But these animals must do what they were born to do. Otherwise, what is the purpose of saving them? You must appreciate animals for what they are, not what you make them to be.

To me, birds are incredibly wonderful. They are the only animal found in every part of our environment. A coyote may be confined to one ecological area for its whole life, but many birds migrate thousands upon thousands of miles. To see a bird fly is marvelous.

I grew up with everything except hoofed animals — cats, dogs, birds, turtles, rabbits, guinea pigs. My brother and I were forever finding things all over the hills. My mother was just as bad as I was. She was a nurturer. I come by it quite honestly. We were just always finding and rescuing stuff and, of course, our cats would bring in things, too. My dad was very tolerant through all of it, but I think he had no choice. It was his daughter, not his son, who brought home the frogs.

As a young adult, I don't think anything interested me other than working and boys. But as soon as I married, I got pets. Then I was ten all over again, only as an adult. I never thought of myself as doing something commercially with the animals; it was always just a fascination and a wish to help them.

Almost eight years ago I began the Wild Bird Care and Rehabilitation Fund. I did it for a long time just as an individual. It is now a nonprofit, registered, charitable trust. I do salvage work with birds only, which means I do not get involved in legislation or anything else. Funding comes from private individuals or organizations. It is somewhat of a problem. When a bird is brought in, the donations are voluntary. No fees are charged. I feel these people are doing such a good thing in just picking them up. To turn around and say you have to pay X number of dollars to get it cared for, when it is not even their animal, seems like emotional blackmail. Some leave $1 or $5. We have kids come in with 80 cents they have collected from all their friends when they have found a little bird who needs care. Their contribution of 80 cents is just as important as anyone else's larger contribution. Legally I could take a salary, but I don't.

We participate in various events but never bring display animals. We do it pictorially. With live animals, I believe what you are doing is prostituting to get attention. Obviously, a live great horned owl is going to draw a crowd. But you are stressing the animal by putting him in an unnatural situation, and that is an animal that you are trying to get back to the wild. But if you can educate, it opens a whole new world for so many people.

There are definite elements in rehabilitation. Imprinting and behavioral modifications can happen, and that needs to be prevented. That is why I avoid constant contact. If they are with you day in and day out, and you talk with them, play with them, and handle them, they learn nothing else but you. They begin to look for people. If they get hungry they will fly to anyone, thinking everybody is their friend. This can be dangerous if the person doesn't know how the bird was raised. The person may think he is being attacked, as in the Alfred Hitchcock film. Unfortunately, those films do a very great disservice.

Raising different birds involves understanding not only what to feed them, but how they normally eat. Subsequently, you must know what kind 27 of habitat they need to go back into and how they need to be socially oriented. There are myriad things you have to do to return them to the wild. The main goal is that they will reproduce the following year. If you release individuals with severely modified behavior, they are sterile. You have done no service at all. You have taken away their birthright, the ability to reproduce.

The only time the work is really bad is during the spring months, when it is horrendous, because the babies require just about constant care until they are on their own. I bring them in at night and put them out in big cages in the backyard during the day until they become acclimated, when they are outside all the time. Once they are totally self-feeding, I leave them alone with food. By that time you have left them alone and you are having no more to do with them other than to provide maintenance. Those birds wouldn't fly to anybody. They are totally wild at that point.

There are some cases that really stay in my mind. We had a little western meadowlark come in from Fish and Wildlife with a fractured wing. He was small, around 8 inches. I got him almost immediately after the injury, which is wonderful. He was taken to the vet, but they do not make surgical intramedullary pins small enough for birds. Our veterinarian had to pin the wing with a surgical wire. We weren't sure whether the bird would fly. Anytime you pin a wing you have to go through the shoulder joint or an elbow joint to line those bones up. Well, there is always a risk, both with the surgery and the anesthesia. But the bird had no choice: It was either euthanize him, have him crippled, or try the surgery. We always want to try because if you get it right and the bird comes through, it may fly again.

Well, we called him Marvel the meadowlark because he not only went through the surgery tremendously well, but he eventually sang his regular songs and flew. I have slides of him being released. It was wonderful, especially when he sat on this doggone fence post and carried on, singing just like he had always been there.

Now there are some birds that can't be released. For years I have had a great horned owl. He was found flying around a neighborhood, landing on people. He was young, starving to death, and had no idea what mice were. He was permanently imprinted on people by someone who had tried to raise him. That is tragic. And you can't place cosmetically injured birds. If the public has trouble looking at humans who are severely disabled or handicapped, you can imagine how it is with animals.

Also, the bird must be able to live with its disability. Some animals cannot, and to try and keep them alive is criminal. A bird like a mockingbird, for example, runs with one foot in front of the other. They don't hop. So if one were to lose a leg, it couldn't compensate. Also, they stop and spread their wings when they are looking for food. A one-legged mockingbird would have to balance with the wing, and would start damaging the end and eventually the joint on that side. On the other hand, we have had little sparrows or small birds that do hop, and though they have lost a foot they do fine. So, what is okay with one bird to live with is not necessarily okay for another. You have to put yourself in their feathers.

I have seen terrible things. One year we had a mature red-tailed hawk with its whole wing shattered. Somebody had shot him. The man who brought in the bird saw it shot, but he needed to get it to help and he didn't know where the people disappeared to. That bird lost his wing totally. Here was a healthy normal bird that was no longer able to fly or do anything normal again because of someone's indiscriminate act. That is a tragedy. The bird had to be euthanized. Its wing was so badly mangled that even to amputate you would have had to take it so far up at the shoulder it would have had nothing for balance.

We are in *their* backyard. We are the trespassers, not the other way around. We have invaded their land. This is what people have to realize. People call me and complain constantly about mockingbirds during the spring — they are dive-bombing the cat or they can't use their patio. Well, I do try to be polite and diplomatic and explain the behavior of the bird: "It is nesting. They won't behave that way any other time, really, except if they have young in the area. The babies totally depend on them not only for support but for protection." People should realize that humans can have children anytime. These animals have only one chance. I try to help people with the problem. For example, whenever they come out they can bring raisins to feed the birds. What really bothers me is that there is so little tolerance for something so minor in the person's life as a bird dive-bombing. Yet for the birds, if they don't reproduce they will not survive.

People should enjoy these animals for exactly what they are: free. That is what makes them unique.

Sloths — Marvelous and Misunderstood

Dennis Meritt, forty-two, has been assistant director of the Lincoln Park Zoo in Chicago for the past seven years. Formerly curator of mammals, he is an expert on sloths. Because of his interest, research, and collecting efforts during the past ten years, the zoo has thirty sloths of two species, the Hoffman's two-toed sloth and the common two-toed sloth. A third of the sloths were born at Lincoln Park.

Dennis commutes to the zoo from his home 13 miles away, where he lives with 29 his wife, Gail, and his daughters, Laura, nineteen, and Jill, sixteen, both of whom share his interest in animals. He always arrives at the zoo before it opens so he can walk through it before the visitors arrive.

DENNIS MERITT: My interest in sloths extends to their whole order, the Edentates, which includes anteaters and armadillos. These animals have been in zoos for years, but no one seemed to be doing anything serious with them beyond simple exhibition. In a sense, they were animals that had been forgotten.

Even before I got into zoo work I thought they were intriguing. All the members of the order have physical and anatomical peculiarities, and are atypical in nearly everything they do. My interest must go back twenty years. When I was about fifteen I used to read anything I could get my hands on that was animal-oriented. Then I refined my interests to the New World tropics, Central and South America, and finally began to concentrate on Edentates. I visited them in zoos, wrote to zoo people about them, and asked about their diets and husbandry. But it was impossible for me to keep any until I got into a zoo setting at Lincoln Park.

The first wild animal I can remember having was my pet alligator, that lived in my bedroom. In fact, he is still alive in the Syracuse zoo. That animal has got to be twenty-five years old now, and it started out as a little hatchling from Louisiana! After I married, I kept hedgehogs and woolly opossums at home. Our daughters have spent nearly their entire life with animals.

In 1967 I had the opportunity to start working with animals at the Lincoln Park Zoo. I had just a few college courses left and petitioned to finish my program at the University of Rochester. I was the oldest graduate in my class. Then, since I have been a perpetual student anyway, I went on to graduate school and finished my master's in biology. Now I have hours toward a Ph.D., and it's a question of finishing that up on a part-time basis as time and finances allow. And I will probably do the thesis on sloths.

I have had a great deal of involvement with the tropics over the last five years. It's easier, and in a sense more ego-building, to work with animals that are larger and more popular, like lions and elephants. But my philosophy was that no animal is too insignificant. And we have this special obligation to sloths because we have put them in a captive environment, and we should gather as much information to gain an understanding of what makes them tick.

I remember my first trip to the Republic of Panama. I had an idea of the animal's behavior from captivity, but actually seeing the first sloth in the forest was an excitement that I will never forget: seeing it upside down in the trees; seeing it moving; seeing it carrying out its life as a free and wild animal. It was, however, particularly satisfying to realize that the only differ-

30

ence was in the setting. I felt very good when I knew that we had duplicated the physical, physiological, and psychological needs in the captive environment.

To study sloths you go to the tropics in the dry season, between January and April, when the leaf loss is at its highest. The rain forest world of the sloths is very beautiful, and it's something that won't be there for very long. The wilds as we know them are shrinking at an alarming rate. It's quiet; it's green; it's warm. All the day-to-day distractions are gone. I find I have an ability to hear very subtle things and see subtle motions. In the broadest sense for me, it's a form of escapism not unlike a religious experience.

You quietly sit and watch the animals with binoculars or a spotting scope. To locate them in the trees is a job in itself, because when they are sleeping they look remarkably like termite mounds or hornet nests. They are kind of curled up, motionless, clinging to the trunk of the tree. You keep a detailed, round-the-clock record on them. In the typical day of a sloth, there are about twenty-two hours of inactivity and about two hours of activity, which is primarily concerned with food getting, breeding, and elimination. I think they are best called nocturnal, but in reality their cycle is centered around dusk and dawn. For these dark observation times you use a night scope, pick moonlit nights, and mark the animals with fluorescent tape. Some radio transmitting work has been done, but all that gives you is position, not behavior.

Normally, they live high in the trees, at 80 to 100 feet. The only time they come to the ground is to eliminate. Unlike other mammals, they eliminate at five-to-twelve-day intervals. This is a wonderful adaptive physiological mechanism to keep them secure. The sloth is painstakingly slow in coming down the tree and going back up. The process of elimination is itself over in just a matter of minutes. The great care that is shown in descending to the ground is because the sloth is really vulnerable at that time to predators like jaguars, ocelots, and mountain lions.

Monkeys, who also live in the tropics, just urinate and defecate from the heights of the trees. Sloths are marvelous and essential elements in the ecology of the rain forest because by coming down the tree to eliminate, they are recycling the nutrients. They eat tree buds, leaves, shoots, and flowers, and when it has been processed, they take it back to the base of the same tree. The rain forest is totally dependent on complete nutrient cycling because there isn't any topsoil. Sloths keep giving back the essential nutrients in a form that is immediately usable. In a sense, the niche they fill in the rain forest is that of an upside-down antelope. When you think about it, the expanse of foliage in the treetops is like the African grasslands.

One of their primary predators is the giant Harpy eagle. It can actually grasp sloths on the wing and rip them out of the trees. It takes incredible strength on the part of the bird. And as somebody who has tried to pry sloths out of trees, I know it is next to impossible for a human being to conceive the strength of their grasp. In fact, the usual way of catching them is to cut off the branch they are clinging to. Their claws grasp like a vice grip. In the two-toed sloth, there are two elongated claws on the forelimbs and three in the rear. They kind of fold at the hinge joint so the base of the claw and the pad come together tightly.

Sloths simply don't move around much. They may shift body positions, but in terms of moving from place to place, they don't do a whole lot of it. It is just like watching a slow-motion movie. The sloth moves from point to point and always makes sure it has a firm grip on a branch or limb. It tests, putting its weight on the branch to make sure it is secure, and it never lets go with its hind feet until it knows a branch will bear its weight. This is a good adaptation to moving upside down through treetops, where there is a constant risk of falling. The movements are very subtle, slow, soundless, and deliberate. There is no need for them to hurry in their environment, and everything they need is within easy reach.

If you tried to assume the normal position of a sloth, you would be lying on your back and looking forward. It is an incredible upside-down view of the world. And a sloth's need for orientation is different from other mammals'. We need to have identifiable objects, but to a tree-dweller like a sloth, it really doesn't make any difference because everything up there is the same. Sometimes I climb adjacent trees and just sit at the same level and watch the animal. I have never had a sloth approach me. I have been quite close to them and they just didn't pay any attention at all; perhaps it doesn't register.

I have never seen aggregations of sloths in nature. They are very much loners. Their home range appears to be very limited, for they spend an entire lifetime within the span of a few thousand square yards. As long as food is available, they stay. They meet each other in the treetops and do not compete with each other or the other tropical mammals.

There is no known elaborate courtship ritual. If males and females meet and are receptive there is breeding. They put their tongues in and out as a sort of greeting gesture that makes a kind of smacking noise. The sound is loud enough to be heard for 15 or 20 feet. Then they touch and lick noses while sniffing about each other's head and neck. They spend hours rather than days together, then go their separate ways. And this is all done in their usual upside-down position.

According to my observations, they have a single offspring, born fully haired with its eyes and ears open. The youngster spends the first four or five months right side up, clinging to the mother's abdomen. As it becomes older, it begins to investigate its surroundings. One of the many long-standing misconceptions about sloths is that if a baby sloth wanders off from its   33 mother, it ceases to exist in her eyes. In fact, the mother is very tender and solicitous of her young. She immediately responds to the bleating sound of a lost baby by waking up, looking in the direction of the sound, moving to it, and making her body available so it can crawl back on. She licks and grooms it and eats its droppings. Eventually the young begins to sleep and rest upside down alongside the mother, with whom it lives in close association for up to a year. They live about ten years, so a female could produce a maximum of four babies in her lifetime.

Over the past ten to twelve years I have gotten to know some of our captive animals as individuals. One I will always remember is Susy, the first female we had in captivity at Lincoln Park. She was exceptionally gentle and exceptionally large. When she produced her first baby, it was like sharing the birth with her. I spent essentially every waking hour watching mother and baby. Both Susy and her daughter, who was named Susy's Baby, are still in the colony. The personality of the child is essentially that of the mother, calm, gentle, very tolerant, and inquisitive.

My main goal has been to see sloths become self-sustaining populations in captivity so we would no longer need to go back to nature to collect them. In the Lincoln Park Zoo, we reached that point five years ago. We have even been able to take offspring from our colony and loan them for breeding to other zoos across the country and elsewhere.

Sloths have been around for millions of years essentially unchanged. They are in a sense in a perfect environment, with everything at their finger-tips, with very few needs or dangers of predation. On the other hand, they are totally dependent on the rain forest for their existence. They even need an exact range of temperature or they go into a torpor. Over the course of my many years of work with sloths in Latin America, I have had animals that I knew well disappear with changes as simple as the selective cutting of hardwoods out of their habitat. These were, in a sense, grandfather or grand-mother trees that had been there for centuries. The animals depended on them totally for food and shelter, and the new growth wasn't sufficient. The destruction is almost sacrilegious. And I just know about this one type of animal. But what about the dozens, perhaps hundreds, of others that share the same ecological niche with sloths? What happens to those lesser forms, even less obvious than sloths?

A Trailer Full of Tortoises

Amid the asphalt, Astroturf, and plaster lawn ornaments of a mobile home park south of San Diego near the Mexican border, the 30-by-70-foot lot of Chip and Carol Wallace looks like an oasis, a small Garden of Eden. Lush bromeliads, juniper bushes, fuchsias, succulents, cactuses, and jungle orchids, as well as miniature lawns and a small fancy goldfish pond, surround the immaculate white trailer. The trailer houses Chip, Carol, her eighteen-year-old daughter, Kris, a hundred and twenty-five tortoises, several tortoise eggs, three cats, iguanas, and three monitor lizards. The 12-by-15-foot living room is dominated by the cages, specially constructed glass-fronted terrariums with stained and varnished wood lining the walls. Outside, the small yard is meticulously organized. One 10-by-30-foot section contains small pools for the semi-aquatic and terrestrial tortoises, and the monitors. Along the back and right-hand side of the trailer is a 20-by-40-foot area for the largest tortoises, some weighing more than 30 pounds.

A few years ago, before the monitor lizards where allowed to live free inside the trailer, the two cats, Smokey and Icarus, would play with Ben, a hooded rat. The rat would race from bedroom to living room with the cats in pursuit. A pretend scuffle would then ensue, followed by the cats racing back toward the bedroom with the rat chasing them.

Once, after I had sustained a traumatic death in my family, several of my own tortoises became uncharacteristically sick and one died. Without mentioning my personal loss, I asked Chip and Carol to check the animals to determine what the problem was. Now, the face of a tortoise seems expressionless, since its leather skin is stretched tautly over its parrot-billed, toothless skull. Only the lidded eyes have soft tissue around them. Yet that evening Carol called and said, "I feel I can ask you this since you are a friend. Has something bad happened? All the tortoises look so sad. They all walked over to me with such mournful expressions. I think they are picking it up from you, and that's why they are getting sick."

CAROL WALLACE: I was born in Alhambra, California, thirty-nine years ago, but my parents moved to Denver, Colorado, when I was less than a year old. I was a resident of Colorado until 1969, when I moved to California with my first husband and my daughter, who was five years old.

Basically I've always been a repressed animal nut. My parents allowed a cat or two, and Daddy always let me take care of injured birds. They're gentle, kind people who like wildlife in the wild, but not in the home with you. I am an only child. (I had a brother who died as an infant when I was three.) My father worked with the Bureau of Reclamation, and my mother 35

was a ninth-grade teacher. Every summer she'd tear down the 20-gallon tropical fish aquarium at school and bring it home.

I married at nineteen, and my first husband loved the outdoors and wildlife. But he was an avid gun nut and a hunter. This bothered me very, very much. My father didn't hunt because he had had a boyhood experience when his father took him hunting and shot a deer. She didn't die right away, and when they went up to her, Daddy looked at the big, brown eyes that he said were actually crying, and he threw the gun down and never hunted again.

I met my second husband, Chip, at a local pet shop. He was basically a hippie, a free spirit — I was an executive secretary in the hotel business, and I had such a life of plastic people and plastic events that I thought, here was a real person. He'd been into reptiles for many years, and he had some water turtles at the time. I had specialized in goldfish.

When Chip moved in he had his big tortoise, G.C. — for *Geochelona carbonaria*, the Latin name for redfoot. Then I bought him a baby leopard tortoise, Tiffany. Then we got a desert tortoise that was found wandering down the street, then two baby redfeet and a box turtle. I would consider Tiffany's first problem as our turning point from amateurs to more concerned hobbyists. When I got her she would only eat lettuce. Then one day she ate dog food and gorged herself to the point where she became constipated and lost control of her hind feet. My first "cure" was when she vomited the excess food after I gave her oral B-12.

We learned fast and we learned under fire. We realized, "My God, this is virgin territory. We don't have anybody to ask, so we'd better get it together quickly." We began to try to rescue animals that appeared beyond help. One positive thing has been that for every death we learned something. We eventually learned to do autopsies ourselves. For example, a Herman's tortoise, who had a nasty habit of climbing and falling, died for no apparent reason. On autopsy we found she was very laden with eggs; when she fell, they had pushed against and ruptured her bladder. It broke my heart because she was simply a beauty.

My knowledge has come from being a veterinary health technician. For a period of time I even planned on becoming a vet. The rest I've learned by attending as many seminars and lectures as I can. And we have a vast library of medical textbooks, from basic biology to advanced parasitology.

Now we have approximately a hundred and twenty-five tortoises — thirty-two species and subspecies, with five protected and endangered — as well as eight large lizards, and our survival rate is very high. We have the Ceylon Star tortoise and the South American Redfoot. We specialize in the genus Testudo, including Horsfield's tortoise, Hermann's tortoise, Mediterranean Spur-Thigh, true Greek or Marginated tortoise, and the dwarf Greek 37 from Israel, and we have just had three successful hatchings of eleven Mediterranean Spur-Thighs.

We now have an African Pancake tortoise that's been living in captivity for eighteen months, a beautiful group of North American Wood turtles, a protected species of which we have a breeding colony. We have the American Bog turtle or Mullenburg's turtle, which is endangered. We have four subspecies of American box turtles: Eastern, Western, Gulf Coast, and Florida. We have the Elongated tortoise from Southeast Asia; the *Cyclemys mouhoti, Cuora trifasciata,* and *Coura flavomarginata* from China; and the Spiny Hill turtle from Southeast Asia. On a cooperative breeding program with two zoos, we now possess the Burmese Brown tortoise.

The most memorable thing was the hatching of our very first captive-born baby — actually to see the pipping of the egg, to see life emerge, was the realization that this is the sum total of why you do it all: why the money outlay, why the tears, why the long hours — this is what it's all about. She was a Mediterranean Spur-Thigh from a captive-raised mother and a wild father.

I have a tragic memory of a little pancake tortoise named Penelope. Pancake tortoises are difficult in captivity. She was about four and a half inches — a fine, feisty little thing. When we sat down in the tortoise yard, she would climb up on us and sit with us. She just liked people. One time she climbed out of her enclosure, got lost, and was gone nine days when something prompted me to look out the window. There she was, heading home. So I ran out and got her. But she died during her first winter due to extensive internal parasite damage. All my force-feeding couldn't save her. It's a very sad memory for me.

Charlotte was our very first big lizard. As an adult iguana, she was found running loose locally and brought to a pet shop about six years ago. She was wild, unsalable merchandise. She had lost half her tail, and her nose was a scabby mess from being rubbed on the wires of the cage. The pet shop owner gave her to us. Now she's a very beloved, gentle animal. The first night we brought her home we didn't know what to do with her, so I put a blanket over the top of one of the tortoise cages, where there was a fair amount of heat. Charlotte has slept there every night since. She is such a creature of habit. In the morning she wakes up and crawls off her bed onto the floor. She marches through the front room, climbs up on our couch, which backs against a large open window facing south, looks out, and suns for most of the morning. Then, after her regular meal of fresh fruits and vegetables, she walks to the cat food bowl in the kitchen, where she snacks on dry cat food and drinks water. Then she does one of two things — goes down the hall to get into the bathtub, where she has been housebroken, or goes to Kris's bed, where there is a down comforter that she likes to get under.

But she's an old iguana now, and we're beginning to see the changes. She looks thin — her legs look thin, her ribs are showing a little, her face has changed. Her eyes are a little more hollow, her reflexes are slower, and she spends more time in the sun than she used to. Recently I have been taking her out to sun every day. I want her to have optimal treatment. We have taken in two new young iguanas, Project Wildlife rescues — Romulus and Rima. But Charlotte is territorial, and we have to keep them separated. I run the iguanas in shifts. When Charlotte goes out, Romulus and Rima come in. Fifteen years is supposedly the record for an iguana in captivity. We've had Charlotte for seven, and she was an adult when we got her.

Our first monitor lizard was a pet shop rescue case, a baby water monitor. I had heard they were very dangerous. Well, this one was half dead when Chip brought it home. It had the most beautiful eyes, gentle and soft. It was pitiful, starving and weak. I thought, "This animal's no killer. This animal's in trouble." So we force-fed it and rehabilitated it before returning it to the pet shop.

Then we got Beauty, a 6½-foot water monitor. She weighed less than 10 pounds and had a severe infection on her right forearm where she had been roped and restrained. The foot was so infected that it was swollen to three times its size. When I picked her up at the pet shop, I wrapped her in a blanket and cuddled her under my coat in the car to keep her warm. She had severe tapeworm and pneumonia. We worked at force-feeding and medicating her almost for four months. She is now a 35-pound healthy, well-adjusted animal.

We got a reputation for our success with monitors. Jake was the most special one. He was on a breeding loan and was, in the beginning, one of the nastiest lizards we ever brought into the house. He was over 7 feet long. He is the only monitor that has hurt me and that was totally accidental. I was hand-feeding him a rat, and he saw my moving hand and bit my thumb. Now I have no feeling in the thumb, but I can use it.

We loved Jake very dearly. He eventually became a very responsive, beautiful animal. He loved people. He was one of the animals that wanted you for the sake of you. It was not uncommon to find him in the afternoon taking naps with my daughter. He knew his name. He could go outside and push open the door to come back in. He loved to swim in the bathtub and almost got to the point where he was learning to turn the faucets on by himself. But because of his early improper care, he had absolutely no resistance to disease. He contracted a very virulent case of streptococcus, staph, and proteus infection, and in spite of extreme cooperation from a number of vets, he died the day after Christmas. I still cry just thinking about him. For a long time I couldn't have his picture up because of the memories.

People will ask what you can do with a tortoise. It's too bad that they are not more attuned to the incredible things they are capable of. They must be seen as individuals. When a new arrival comes, the most important thing is to get to know the animal. That way I know what is normal and what isn't. Usually it takes me about three days. Seeing them as individuals is what I'm striving for.

I have a daily routine, seven days a week, three hundred and sixty-five days a year: I am up and moving at least by 6:30 A.M., at which time I start by setting up the tortoise food and turning on the lights of the cages of those that are inside feeders. I make my husband's lunch and have a cup of coffee. Then I fill the bathtub full of warm water for Beauty and Charlotte. I carry Charlotte into the bathroom; Beauty walks in by herself. Both lizards poop in the tub and then come out and begin their own daily routine of driving me absolutely nuts by meandering around for several hours. By then it is seven-thirty or a quarter to eight, and I set up the food for the outside tortoises. Then I begin to hand-carry our approximately hundred animals from the indoor cages. I probably haul at least 150 to 200 pounds twice a day — out and in. All the tortoises are put into their respective yards, and I spend time watching to make sure each is all right.

The only problem I find with the small yard is with personalities. Some are feisty and some are very shy. During the day I have to check every thirty minutes or so to make sure nobody is knocked upside down by a more aggressive one. Sometimes they show me when they are ready to relate. An aminal that has been here X number of months and has obviously been treated on an equal basis will one day make an overt gesture to come up to me in the yard. If I extend my hand they will come up to investigate. If I can reach a mutually unafraid relationship with an animal, I feel I have achieved what I want.

In midmorning I go in and clean cages. In winter, when the animals are confined, they are sometimes cleaned up to three times a day. I wrap up at approximately noon. That leaves me some time to put on my other hat, which is housewife/mother, and to do any personal activities I may care to indulge in. By late afternoon my husband and daughter are home and help me bring the animals in. I fix dinner. Then I do telephone work with Project Wildlife and the Turtle and Tortoise Society and do my medicating, injecting, force-feeding, and surgery. I do try to watch educational television once in a while to relax.

My worst problems are in the winter. The majority of our animals don't hibernate. If I'm lucky and it doesn't rain, I get everybody out by, say, noon. Then they can get a few hours of sun. But if all of a sudden it clouds up or the winds start, it's a hustle to get everybody back in. I get depressed that time of year and worry about illness or death.

Our place always seems to get a little shabbier. I keep wanting to replace the floor covering and furniture, but there always seems to be some animal that has to be paid for or medicines and animal food to be bought. We only have one couch, one chair, and one small coffee table left in the front room. The rest is cages. We have cages literally stacked and then separated into compartments according to heat requirements.

The incubators are in the closet that was originally designed to house the washer and drier combo. Each can handle forty-five to fifty tortoise eggs. They are homemade from Styrofoam ice chests. The most important part of hatching tortoise eggs is patience. It takes between sixty and ninety days with some, and up to a hundred and twenty with others. An egg that is not fertile can be candled in fifteen to thirty days and will still be quite clear.

On a light note, we have a car that is going to fall apart any day now, and we have this floor covering fund to get rid of the carpet that literally has holes like Swiss cheese; but we have just agreed to buy five Horsefield tortoises, and a lady from Florida called with a female marginated tortoise for $100. We don't have a pot to piss in, but since these animals are necessary to our breeding program, I have $300 worth of tortoises coming in less than a month!

CHIP WALLACE: There are lots of things we want to do, and we probably could if we were anyplace but here, but we're in a mobile home park and are limited in space.

I was born in Pasadena, California, in 1946. My dad didn't care about animals; he once made a statement that if all the animals fell off the face of the earth he could care less. My mother, I think, had a rat as a pet and a hamster and either a dog or a cat — but having seven children plus being a schoolteacher kept her busy. But we were allowed to bring small animals home as long as we took care of them. Originally I was into birds — pigeons, doves, chickens, ducks, owls, hawks, and sparrows. I also had pet rats. At fourteen I started out with a few turtles and an American alligator because they were legal then, but I was terrified of snakes.

By the time I moved to San Diego I was interested in snakes, but got involved in starting the San Diego Turtle and Tortoise Society. Ever since then, turtles and tortoises are basically my life. Just prior to my meeting Carol I went for a whole year almost without pets because I felt they were literally consuming me. But I had kept G.C., a redfoot tortoise, for over ten years, and from the time he was a 5-inch youngster.

Tortoises are a gentle, curious animal by nature. They are fascinating in the sense that they are enclosed within a shell and can't climb. Their life is awkward, but they have managed. They are individuals. Some have no smarts at all, like some humans have no smarts. Yet there are some that know their names.

We try to make life easy for them, even though it's a small environment. We try to make them feel secure and fulfill their life requirements. Tortoises are not dependent animals like a cat or a dog. An insecure person couldn't relate to a tortoise, because they don't satisfy a person's ego.

An Orphanage for Wolves

Ben Bok and Donna, his wife, live in the high desert northeast of Los Angeles. There, at an altitude of 5000 feet, the land looks surreal, with huge boulder outcroppings, stands of tall, palmlike joshua trees, and sparse forests of rugged black pines in the most mountainous areas. The winter brings freezing temperatures and snow.

Their 40-acre property overlooking the desert floor has three chain link–fenced pens, each enclosing several acres on which Ben keeps his wolves — three packs of over ten animals each.

The Boks live in a trailer surrounded by a weathered wooden porch. They have neither electricity nor a telephone. As dusk deepens, they light glass kerosene lanterns. At the front of the trailer, about fifteen short-haired sled dogs are chained beside large doghouses. To the side of the trailer are enclosures for over twenty mothers and puppies of the same breed. Ben's hobby is sled dog racing, and he has a beautiful, handmade, wood and rawhide racing sled that his teams pull over dirt roads.

When one drives up to the house in the evenings, the unforgettable howling songs of the wolves reverberate through the hills.

BEN BOK: I don't believe anyone in the family has a history of having anything much to do with animals. I was born in Pennsylvania and had one brother and one sister. We moved around a lot. My parents were divorced when I was about seven. My father was a judge in Philadelphia; my stepfather was a plastic surgeon. We always lived in sort of remote places, so the three of us, my brother, sister, and I, played together constantly. My sister is married to a veterinarian. My brother is president of Harvard.

My wife and I were kind of lost souls wandering around. (I had been married before and divorced.) We have been married almost twenty years now. We always seemed to have animals, first cats, then sled dogs.

Before I had a wolf I was pretty much a wolf expert. I had read all the literature and I had it all at my fingertips. Then I got a wolf and was an even bigger expert. Now that I have thirty-three, I don't know anything about wolves at all. Everything I thought I knew has been overturned by these animals. They just don't read books.

About twelve years ago, I became interested in wolves and bought a wolf pup. I had been involved in sled dog racing, and one thing led to another. I had always enjoyed wild animals, anyway. Wolves are very easy to get. It's a very profitable business to raise them. They sell from $500 to $1000 apiece for a little two-week-old pup, with no guarantee it will live. You have to feed it on a bottle. Unless you start them at a very early age they become suspicious of people.

We wanted a small female and we went to a guy who had some pups. He was leasing wolves to a zoo for breeding. As part of the lease he got some puppies every year and sold them. That fellow is no longer in the business. He needed it economically and gave it up as soon as he could. The people are always calling and begging you to take the wolves back within two years. It doesn't work out very well.

We named our pup Houlun, the name of Genghis Khan's mother. We began to meet other people who had wolves, and we found that every one of them wanted to get rid of them. There was a vast demand for an orphanage for wolves. We had the money and decided to start a place. By then we had lived with Houlun in the house and a bunch of sled dogs out in the yard for almost two years. It was unbelievable. You had to put a clothespin on your nose. You can't house-train a wolf. They go where they go and move on. We had to lock her in the house at night so she wouldn't howl under the windows of the apartment house next door every time a siren went by. The rugs were torn to shreds. There was no possibility of any kind of curtains or blinds or drapes or anything. Certainly no overstuffed furniture.

In those days I was a postman. One day I was walking my route. One of the people on my route had thrown out an overstuffed chair and a couch. They weren't good enough for him, but they looked good to a wolf owner. So after work I went over with my truck and hauled them home. Then I sat there in the living room and watched while Houlun took just about twenty minutes to take each piece down to the bare frame. No way you could stop her. She ate our records, but that didn't matter because she had eaten the hi-fi already. The thermostat was torn off the wall. She did anything she wanted. They simply don't understand if you get upset about it.

You can train a dog because a dog thinks you are God. When God gets upset, that is a terrible thing. If he poops on the floor, he doesn't really know that is wrong, but he knows it gets you upset so he quits doing it. That is the way you train a dog. You don't have to hit a dog or anything, just show him you are upset. But with a wolf you are an equal. The way a wolf looks at you is like this. It feels sorry that you have this emotional problem about not wanting your furniture eaten, but it is not his problem. He knows that it is right to rip up the furniture if he wants to do it. It's your problem, not his.

The only way to keep a wolf is to enjoy it for strictly what it is and to be willing to put yourself out tremendously for it. A wolf is not a pet. You can't expect anything of it. It's a wild animal. It's the product of millions of years of evolution and was meant to survive in a very rough environment. Almost all of the animals we have are here because their owners tried to do the impossible: They tried to train the instincts out of the wolf. Yet instinct is as much a part of the animal as its heart or its lungs.

Most people who get a wolf have the idea that they are going to be walking down the street with this 150-pound animal on a leash. They picture that people will be standing there saying, "Wow!" First, there is no wolf that size in the world! Second, it doesn't work out because the damn thing completely freaks out in public. It literally climbs the walls. You can't possibly lead it around. Then you invite your friends over to see your pet wolf, and it hides in the farthest corner of the backyard. You worry about neighbor children climbing the fence and getting torn up, or the wolf getting out and running about the neighborhood killing dogs. They are lethal. There are really very few satisfactions to owning one as a pet. People just get sick of them.

I have often thought that if I were younger (and I have tried to talk younger people into it), I would get a good book on dogs, do a little study, and then go to the pound, where you can pick up just about anything, or maybe even spend some money and get a purebred. Then I would breed a super Disney wolf with no wolf blood in it at all! Breed 150 pounds of pure Disney wolf and sell it. You could just write up a little brochure that didn't exactly lie, but that would make it easy for people to kid themselves. You could make a million dollars selling these because that is what the people want: a thing that looks like a wolf but is a dog.

For many years there was all this denigration of the wolf. People talked about what a horrible animal it was and how they all ought to be shot. But then the pendulum swung the other way and there have been all these books, lectures, and TV specials about what a nice guy the wolf is. It has gone too far in that direction. All the stuff about the social life of the wolf is obviously interesting information. But there is just not enough talk about the other 50 percent of a wolf, which is equally important. In fact, the animal is a savage, aggressive killer. It walks around in the woods and attacks moose, for example. There are plenty of people up in Alaska who will tell you that a moose is the most dangerous mammal in North America. They will say it is worse than a grizzly bear. Whether they are right or not, it is no sweetheart to mess around with if you have to attack it with your mouth. In a pack a wolf can kill a moose. It is capable not only physically, but emotionally, of doing this sort of thing. That is what a wolf is. That is an equal part of it. If you are just thinking about wolves or watching TV stories about wolves, it isn't too necessary to think about that side. But when you are dealing with 45 them as we do every day, eyeball to eyeball, you have to bear in mind that other side.

You may not believe this, but I have seen a wolf bite nine-gauge chain link fencing like wet spaghetti, tear a hole in it, and walk through. It is a spooky thing to see. One of the things that happens when people get a cute and cuddly two-month-old pup is that it gets out in the kitchen and grabs a can of cat food, and the thing looks like an eighteen-wheel truck had run over it. It is flat and perforated and he is sucking the juice out through the holes. I have seen a three-month-old pup pick up a 60-pound sack of Ready-Mix concrete, which was the second one from the top in a stack, and walk off with it. If you have ever had your hands on a lion, you know it is all just steel cables and concrete. A wolf is hard like that. The only thing I could compare it to is my sled dogs when they are in tiptop racing condition, when they have run 400 or 500 miles in training and are tough, tough. But a wolf feels that way normally.

It is a lot of work around here, but it is really worthwhile. It is a charity essentially. Some people stop in the street when a little dog has been hit by a car, take it to the vet, pay the bills, and find a home for it. Other people send care packages. This is what we do. We haven't taken a new wolf in for a long time now. There is just no room at the inn. We have one very old one that will die almost any day now. She is a famous old wolf. She was called "a 200-pound wolf." Her actual weight is about 105 pounds. I have seen only two wolves bigger than she is. In truth, females average around 65 pounds and the males around 78 pounds.

I like wolves for several reasons, the first being that they are relatively easy to get along with. I can best express it by saying that I don't like dealing with big cats because they are social climbers like humans. You have to be on the dominance kick with big cats. If you are down a little bit (had a fight with your wife or a cold in the head or anything else), they will spot your weakness instantly. Then you will be in greater danger because they always want to be top cat. A wolf is not that way. A wolf is a social animal who just wants to belong in the group. He doesn't much care where for the most part. He doesn't aspire to leadership. Therefore, if you handle a problem with a wolf once or twice, you have it solved for all time. You don't have to constantly be lording it over him and reasserting dominance. I like that because I just don't like constantly coming on strong. I can do it if I have to, but it isn't a big fun thing to me. For certain wolves my body language must say, "I am King Kong. You mess with me and I will stomp you into the earth." There are other wolves that I never look at directly when in their pen. If I want to know their physical condition, I check after I come out of the pen. Then there are one or two wolves that I go into the pen with and my body

language says, "I am just a little white rabbit with a twinky nose and pink eyes and I wouldn't bother anybody." Those animals will jump you if you challenge them in any way. You have to play whatever is called for by their personality. But this constant macho thing with a cat is tiresome for me.

The other thing I like about a wolf is that it is an equal. I love going on a walk with a wolf, for example, because it is like going with a friend. We sit around and we talk about things and it is give and take. I don't like subordinates. I don't like dependencies. But the fact that wolves get along together makes people assume that they are nice, easygoing characters. The way a wolf pack really gets along is the way we would get along if each of us had a .44 magnum pistol on his hip, a real knock 'em down and stomp all over them pistol, and if we were known to have the will to use it. We would treat each other with a lot of respect. So physically I may be capable of knocking you on the floor, but if you say, "Back off, buster," I say, "Yes, ma'am," and I back off. That is how wolves really do get along together. One growls at the other, and the other knows that he has the means to kill and the willingness to fight. They don't growl as a bluff. That is how they get along. It doesn't preclude love, you know. They have a lovely relationship based on mutual respect and love for each other. One of the great sights around here is when they all get together around the leader. They come up from underneath him and tell him how wonderful he is, and how magnificent he is, and how secure it makes them feel to be in his pack. It is a lovely thing to see. But each one is still carrying that pistol. No doubt whatsoever about that.

Their howling is a mysterious thing. You wake up on a cold winter night. You are in long underwear underneath blankets and it is 10 degrees outside. You will hear a wolf howl all by itself. It will then go through the entire rise and fall of the howl for ten seconds or so, and then it fades away. Then you will hear the whole group come in. It is as if Toscanini had dropped his baton: wolves from here and wolves from there, and with no sight communication between. They just come in as if an orchestra conductor had given the signal, and they will go through the whole howl and stop. Then they do it again. You wonder how these animals know. Their timing is just perfect.

My most unforgettable wolf was Angeline. She was a medium-sized black wolf about eight years old when I first met her. She had been boarded in various places all her life. The owners loved her but they were never in a position to keep her, so they left her with us. She had a tough personality, not vicious, but so powerful that it simply would put up with no crap from anybody. When I saw her I thought she was the most gorgeous female, except for one human female, that I have ever seen in my life. When she first saw me she instantly took to me. That night I came home from a meeting and I went up to check on her. And there she was, paws coming out at me 47 through the space in the gate. I thought, "Gee, that seems kind of friendly." Because I had heard all these stories about her, I didn't trust her in the dark, but the next day when I went up, there she was, making all these friendship signs. She was just all over me. It was wonderful. In fact, I finally crouched down and she came up and opened her vast mouth and took my whole throat in her jaws. I wasn't really worried because a wolf is not a faker. If a wolf says, "I am your friend," unless you do something to change it, it is not going to close those jaws on you. So my whole life did not pass in front of my eyes. I did think of some of the bad stories I had heard, but I had faith that she wasn't going to fake me out. She isn't a human. She's a wolf. We got along beautifully, but always with the utmost formality after that first meeting. Just once in a long while she would come around and indicate that patting was permitted. I could pat her on top of the head, but not the back. She would sometimes lie down and I could pat her on the chest, but not the belly. She would go for you if you went to the wrong place. One day she had her paws up on my shoulders and she was licking my face and my mind wandered for a moment. I put my hands down and started running them down her flank. People probably remember that day, because that was the day the sun stood still in the sky and the birds stopped singing all over the world and everything came to a dead halt. It was just like that. Suddenly these cold yellow eyes were looking directly into mine. She didn't raise her lip, or growl, but everything just became very, very still. And I just took my hands away and put them down by my side. And she looked at me a bit longer, then went back to licking. I nearly lost my face that day.

I adored that animal, and I get choked up when I think about her now. She had never been out of a cage in her life until I put her out into one of those big enclosures. I will never forget how she first slunk around by the walls and then went out into the open. She just glowed. She lay down and rolled. She was so happy. It was wonderful.

We had her in with a male for about three months and they seemed to be getting along all right, but there was never any close interaction. One morning I went out and she was dead. I took her to the vet to be autopsied because I didn't know what had killed her. When we got her there, I just broke down and cried. I still get teary just thinking about it. I loved that wolf so much. It was a throat bite. The male had killed her. We put him with four other females under very close supervision. They all tucked their tails and ran. Then we put him with a fifth, and they fell into each other's arms, and they have been happily married ever since. That is a wolf for you.

Our interest here is in ending the pet trade in wolves. I am not sure that I would favor a legislative solution, because I am dead set against coercion of all kinds. But insofar as I can convince people, I take the opportunity to say that wild animals are best left in the wild.

I really don't see a sharp dividing line between species. What I am doing is a charity — the same thing one does when one sends care packages to people. These are orphans who need help. When they come to us, we're the last stop.

Swimming with Humpback Whales

For over a thousand years people have hunted whales for oil, meat, bones, tooth ivory, and ambergris. Baleen was once used in the manufacture of corsets. Today's by-products of whale bodies are margarine, soap, linoleum, synthetic resins, dog food, cattle feed, glue, gelatin, vitamins, hormones, cosmetics, and gut for tennis rackets. In 1946, the International Whaling Commission placed restrictions on whaling, but at present the bowhead, the Atlantic right whale, the Southern right whale, the blue whale, and the humpback are all in danger of extinction.

Dr. Sylvia Earle has been active in whale research, whale conservation, and deep sea diving. In 1971 she emerged from a submersible craft to the sea floor off Oahu, Hawaii. The depth was 1250 feet, making it the deepest solo exploratory dive ever done in the open ocean. Protected by a specially designed suit, she had only a communication tether joining her to the transport vessel. She lives with her mother, her daughters — Liz, twenty-two, Richie, twenty-one, and Gale, fourteen — two dogs, four cats, four geese, spiders, frogs, and turtles.

SYLVIA EARLE: I've always been in love with the ocean and have spent a lot of time there ever since childhood. I was born in a little place called Gibbstown in New Jersey. My mother has a very warm attitude toward all living things, human beings included. She would bring creatures into the house for us to get to know. Mother had training as a nurse, but more important in her attitude was the fact that she grew up on a farm. She developed a basic feeling of appreciation for life and of the importance of the ways different creatures interrelate. For instance, spiders were always respected citizens, and if one was in the way of a human being, it was carefully picked up and taken out to some more hospitable place. Dad was a very gentle person, not one who ever had to prove anything by being brutal, though he used to hunt.

When I was seventeen I took a summer class in marine biology. That's when I first had the opportunity to use a tank and regulator; there were no formal diving courses then. My first scuba instruction consisted of two words: "Breathe naturally." I remember putting on the tank and the big double-hose regulator. My first dive was out in the open sea, and I practically 49

had to be pried out of the water. It was in the Gulf of Mexico, 5 miles offshore, in the middle of a grass bed filled with just so much action — little fish, starfish, sea urchins, crabs. Immediately scuba became a means to an end, not an end in itself. It was glorious, and I really haven't changed in my feelings. It's still glorious.

I'd seen dolphins, the small whales, from the time I was a child. But the first time I saw a whale up close was during an expedition to the Indian Ocean in the mid-sixties. And I saw what most people see of whales when they see them from the surface: I saw a spout. Even though that little puff of vapor was such a small thing, it boggled my mind, knowing that under it was a creature perhaps 40 or 50 feet long. I participated in a total of five cruises in a period of three years, and I tucked away in my mind the thought that it would be fantastic to really get to know whales underwater. But the difficulty in doing so seemed enormous. They are high-seas, open-ocean creatures. They move fast.

How whales live and what they do is relatively new information. Historically, most of what is known about whales has come from dead whales or from pursuing whales in an adversary position. It is only in the last twenty years, and more particularly in the last ten, that people here and there around the world have begun to look at them in their own element. The early whaling records were all written in an atmosphere of death and destruction. It was a war in which human beings were there with harpoons, describing the attacks on whales: for example, whalers as a matter of strategy would wound a baby whale, and the mother would stay by the young one to defend it. They would kill both her and the young one. And there are writings about the loyalty of a female or male to its dying mate.

Our humpback whale project began in 1976 when Roger Payne, who produced the record of the songs of humpback whales, gave a talk in New York at a meeting where I also gave a talk. He was entranced with the underwater approach and was interested in my experience as a diver and a biologist. I was impressed with his many years of experience in studying the songs and behavior of whales. We talked nonstop for five or six hours about our mutual interest in the ocean and whales and about Roger's long-standing interest in trying to get to know humpback whales under water and do a definitive film about them. What he proposed and what we developed in the course of the evening was a collaboration.

The humpback whales tend to be shallow-water whales. They don't seem to go much deeper than 600 feet, unlike sperm whales, which routinely dive to a mile or more in search of their prey, the deep-sea squid. The humpbacks are near-the-surface feeders.

Our study began in Hawaiian waters in early February of 1977, when we came to a group of whales. Whales tend to congregate and to follow migra- 51 tion routes. This group was cavorting around, spouting, flippering. We tried to maintain a respectable distance to avoid disturbing them, but they changed direction and came to us. They went under our boat, turned their enormous bodies sideways, and looked at us. We turned the motor off, and sat there and watched them. Having convinced the National Geographic Society, the World Wildlife Fund, the California Academy of Sciences, and the New York Zoological Society that what we really wanted to do was jump in the water with whales to get to know them on their own terms, we found ourselves at the moment of truth, having to convince ourselves that we really did want to jump into the water with 40-ton creatures 40 feet long. Of course, the hesitation was brief, and we all jumped. But I was impressed with how vulnerable I was. Clearly, they were the masters of that realm. Right away, a 40-foot-long, 40-ton female — I calculate roughly one ton to the foot — came straight for me. I was at her mercy. To think of getting back in the boat was unrealistic. I decided that I might as well relax and savor every moment since it might be my last!

I tried to remember their gentle and nonaggressive image as this freight-train-sized female zoomed toward me. She came almost within touching distance, and at the last possible moment before we would have collided, she simply tilted her flippers and went off to one side and looked at me, rolling her great grapefruit-sized eyes as she went by. Then I relaxed, feeling reasonably sure that she knew exactly what that big body of hers was going to do.

This was confirmed a moment later when she went toward the photographer, Al Giddings, one of the pioneers in creative underwater photography. He had his eye up to the camera and didn't see her coming. She approached with her 15-foot flipper coming clearly at the level of his neck. She surely could have decapitated him. But the flipper just went *whoosh* and up over his head. He felt the wake and looked around. It was a question of whose eyes were bigger, the whale's or Al's, at that stage.

From then on it was like a ballet. For two and a half hours there were five whales just swirling and tumbling around us like otters. It destroyed forever the image that is presented in books (and that I still carried in my head) of whales looking like Greyhound buses or big loaves of bread — stiff, elongated creatures that go along like submarines in one plane. They do take on a horizontal orientation at the surface since their blowhole is on their back, but just before they get to the surface they could be in any position — upside down, standing vertically in the water, or curling and moving in a three-dimensional atmosphere. And it's just wonderful to watch them. They did go out of their way to engage us. They didn't have to do that. I would view it as a kind of curiosity. They could have left at any moment — it was totally up to them — but they stayed and swam around and around us.

The whales sing at all depths, but the greatest number we encountered singing were at depths of between 40 and 60 feet. We wore microphones, and you could hear the song when you were in the water. It was just overwhelming — just like being next to an orchestra. When you're near a singing whale your whole body resonates. It is an enormous sound out of an enormous animal, and in all your air spaces — your chest and your sinuses — you just feel this big vibration. It's beautiful, yet it's so intense that it's almost painful sometimes. The song itself is beautiful to my ears, though some elements are harsh. Sometimes it sounds like a barnyard with a lot of *heehaws* or *waarcks*. But then there are melodious sequences that resemble a flute. Their body is the instrument and the receiver.

The structural information about the song has been evolving. It has been shown by Roger and Katy Payne that the songs change every year, and that during any one season all the whales in a particular ocean are singing the same song at any one moment, but that if you come back the next year the song is different. Whales in different oceans sing different songs. I don't believe that it is simply a courtship thing. There is some evidence that female whales may sing too. Certainly all humpbacks vocalize, using various social sounds as well as the distinctive songs.

We were not very satisfied the first year about the results of getting to know individuals. But we saw a lot of whales and took a lot of pictures. Then we traveled to Alaska, which at that time was believed to be the summer feeding ground for Hawaiian whales. Observation in Alaska was totally from the surface. The ocean visibility was only about 6 inches to 2 feet, but our encounters with the whales were some of my most exciting. It was there that a year earlier a researcher had reported that whales were making circular "nets" out of bubbles and were using them as a tool for capturing krill and fish. This was viewed with great skepticism by people who said that whales don't think. It was also observed that whales sometimes cooperated to make a giant net, as much as a hundred feet across. When they did this, they were reported to make certain consistent sounds.

Watching the whales feed in that icy cold water was in itself an adventure. We found ourselves getting acquainted with the whales and their behavior and feeling increasingly comfortable working around them. They seemed to know where their big bodies were, and even in opaque waters they avoided contact with our small boats. Though there were a few notable exceptions. On one occasion, three whales were feeding side by side. When they do this, they touch and develop a synchrony. They just gulp, gulp, gulp, with synchronous action like the Rockettes. Sometimes as many as six whales can do this sequence. One apparently takes advantage of the water being spilled out of the adjacent whale's mouth. It is beautiful to behold. But as we were watching the three whales, they ducked under our rubber boat and rolled it up on one pontoon. The boat was about 17 feet long, and the whales were three times that long. It was rather frightening. The boat, however, didn't tip over and we were fine.

We've gone back every year to continue documenting and describing whale songs and underwater behavior and identifying individuals. But the most exciting thing happened when I was hundreds of miles away from the nearest whale — in Geneva, attending a meeting. I got a telegram from my daughter that sounded like a James Bond communication: SPOT AND NOTCH-FIN HAVE BEEN SEEN TOGETHER IN HAWAII. I leaped for joy and let out this big whoop, which totally disrupted the meeting. I described what had happened. Two whales named Spot and Notchfin, who spent a lot of time together in Alaska, had just been positively identified in Hawaii. They were both females — maybe mother and daughter or just friends. Now, with our files of photographs, we can document migrations and get some insight into their social systems.

We're beginning to get enough information to have the potential for answering how long a mother stays with her young, and whether "staying together" for a whale means touching distance or calling distance. Is the social group one that is connected by song? Songs travel immense distances. With blue whales, for example, the ultra-low-frequency sounds travel perhaps hundreds of miles and in earlier times maybe even thousands of miles. Now there's so much new noise in the ocean, it's difficult to say how far the songs can be heard. Another problem is that the number of whales has been so depleted.

This work is the beginning of being able to protect whales. The real beginning is, of course, to stop killing them. Then you have to protect feeding and breeding areas. For example, with the humpback whales, the breeding area is in Hawaii, and the feeding area is in Alaska. With that information you have a fair chance of establishing meaningful whale sanctuaries.

I have never questioned the idea that there is a continuum of awareness throughout all creatures, that all life is responsive. It's hard to say what the dividing line between "responsiveness" and "awareness" is. There is no question in my mind that whales are intelligent by our standards, but I think by their standards there are other criteria that would cause us to feel quite humble in terms of decision making and problem solving. But we don't know enough yet to know how to frame the questions. I surely felt an exchange of awareness when the whales looked at me and I at them.

Orangutans at the Dinner Table

Francine Neago, a divorcée in her late forties, lives in a two-bedroom converted garage in the Simi Valley of California with her 93-year-old mother and Bulan, a four-year-old Sumatran orangutan whom she has raised under a special permit for keeping endangered species. Behind the house is Bulan's private trailer, filled with cushions and toys. No people are allowed in it. Outside the trailer is a swing set that Francine constructed for him.

In the early sixties, Beatrice and Alan Gardner taught a female chimpanzee named Washoe to communicate by using the sign language of the deaf. Their work provided an inspiration for Francine, who got Bulan with the intention of expanding the teaching of sign language to yet other primate species. She also hoped that by raising him like a child and by being constantly with him, she could enhance the process of socialization and communication. In the backyard are the chickens, ducks, and rabbits she has raised to expose him to other animals, and he is quite gentle with them. Adu, the German shepherd, was acquired from the pound at the same time as Bulan to provide additional companionship.

Bulan is incorrigibly mischievous. One day when Francine left me alone with him in the dining room, he politely waited on the couch until he was sure she was gone. Then he sprang up, swung open the refrigerator door, pulled out a gallon bottle of wine, uncorked it, gulped a few sips, recorked it, and returned it to its place. When Francine returned he was sitting placidly, just as she had left him.

FRANCINE NEAGO: I was born in Paris and went to a convent school there. My father, a violinist, conductor, and composer, was a Rumanian refugee from the Russian Revolution. He came with the Russian ballet, the music of which he was conducting. My mother is a pianist and painter. She met him in Paris and they fell madly in love at forty. They'd never been married before. My parents both had very artistic temperaments, and they spoiled me. I had an older brother who would boss me around as older brothers do. (I had been a twin, but my twin, a boy, died within three months.) I was the cherished one. My father never wanted a girl, but when I came into this world he adored me, and he adored me right up to the time he died, seven years ago. He is the one who taught me to love animals. He was a very sensitive man, a very artistic man, and like most great artists, extremely close to nature. He loved to take long walks, and we young children didn't like to take long walks, so he would simply drop us at the zoo. He found that I enjoyed the monkey cages more than anything else. He knew where to find 55

me, and he would say that five hours later he could still find me glued to the monkey cage.

I went to nursing school and finished when I was twenty-one. My father was beginning to travel with his orchestra, and I went with him to Cuba and South Africa. Mother always stayed at home. I just loved traveling and I became very international. I speak six languages: French, English, Spanish, Italian, Yugoslavian, and Indonesian. After about seven years of travel I came to America, to New York. There I fell in love with an Indonesian M.D. at Columbia University, where I was taking a course in psychology and he was studying parasitology and immunology.

I said I would go to Indonesia with him, not just because I loved him but because I also wanted to learn what the country was like. I was thirty-one and I had never been to Asia. But when we arrived, his mother had arranged everything, and I was married the very next morning by a Moslem priest. I lived there for sixteen years. He was tolerant and said I could have anything I wanted to make me happy. I started with monkeys before I could even talk the language. I went off with my driver, who would pedal me in a cart drawn by a bicycle. I knew the name of the market, but I pronounced it wrong, so we pedaled around for two hours, and then I couldn't explain that I wanted a monkey. He thought I was talking about chickens. I eventually got to dirty little shacks where they had the most remarkable and rare species of birds and all kinds of animals. Finally I found one little macaque and got him for something like 50 cents and took him home so proudly. He became my first pet.

Three months later the Indonesian revolution occurred. For about twenty-four hours it was a communist state. Then the whole country reacted against the communists. It was a very, very bloody, terrible revolution. This was '63. So I thought, "Who can play with monkeys when humans are being killed." So I founded a hospital in a school building. My husband and I decided we would be on the Moslem side. If the communists won again they would kill us, but at least we'd rather live as Moslems than live as communists. Having decided that, we picked up the soldiers in the street who were Moslems. We couldn't possibly pick up the communists too, because if we did, they would kill each other in the hospital if they had any strength to stand up at all. It was easy to tell the Moslems because they were the ones who had the big, bloody, messy wounds of dumdum bullets, which would explode within the flesh. The nice, clean wounds inflicted by the Moslems on the communists were made by the ceremonial daggers that everybody has.

I saw my macaque every night. I kept him right up to the time I left Indonesia. Then I released him into the woods with his female and his baby. They must be proliferating there now. A year after the revolution, when my 57 last patient walked out, I decided to change the hospital into a prenatal and postnatal clinic with a dispensary so I could give the poor some help. Mother joined me in Indonesia three years after the revolution.

One day when I was sitting in my dining room, just finishing eating, a man came with a little two-and-a-half-year-old orangutan, who looked like a little clown as he walked absolutely straight. He sat at the table and he pointed to the food instead of grabbing it as any monkey would have done. I was just fascinated by him and I said, "Oh, I want him." So I didn't bargain in a country where you absolutely have to bargain. I paid about $200, which was far, far too much. I could have had him for $30 or $40 if I had bargained, but I was so afraid that the man would walk out of that door and I would never see him again. By then I had accumulated twenty monkeys, brought to me by my patients. Yet he was so different. He wasn't a monkey at all. He was a child, a retarded child of some sort. My husband said, "If you want a child, I can give you a child. You don't have to have this kind of an animal as a substitute." I said, "It's not a substitute. I have always preferred monkeys, and if I can have an ape it would be all the more fun."

The maid called him Tuan, which means "sir." Of course she used to call my husband Tuan too. So it was funny because the two Tuans would respond with the same kind of grunt to the name! He ate with us. Then my husband decided he wouldn't eat with us, that he would eat at another time, so Mother and I would eat with the orangutan. This was very awkward, but by the time I had three orangs my husband decided to eat with us again, and they were well behaved enough.

I got the other orangs after Tuan became very sick, about a year and a half after I got him. I suddenly realized how much I was involved emotionally, and I became afraid of losing him. I decided to look for another orangutan in Borneo. I went there by myself and did some jungle tracking, but to no avail. Then I went up to the villages and asked everybody if they had seen one, but nobody had. I'd found all kinds of other animals, among them two more gibbons. But I wanted an orang. And just on the day I was to leave somebody brought one. He was three and a half years old. They had named him Dulu, which means "later." He hadn't been out of the jungle long because he was extremely nervous. Tuan had been very independent. This one was just clingy, clingy, and he gave me a lot of cuddles. He was very bratty and not too intelligent. Tuan was extremely intelligent.

I used to take "my two little boys" to the zoo, and once when I went there I learned that eight of their orangs had gotten struck with polio. Three or four of them had died already, and one of the big females I had always loved was paralyzed. So I immediately said to the director, "If you will allow me to take her home, I will look after her and give her physiotherapy." I took her home. She was fully grown. I was longing for a fully grown creature anyway, and though she was extremely limited in intelligence, she looked after Dulu. But to her, Tuan was the lord and master, although he was only about six then. She recovered almost fully except for a slight limp. She was extremely tame and had a very, very gentle temperament. She'd been locked up in a small cage all her life and I reckoned she was about fifteen years old. It was great, like slowly expanding your family.

But I always felt like a foreigner in Indonesia. More than fourteen of those years I lived with the orangs. I have so many memories. They would sleep in their own beds, with their blankets. By the time Tuan was seven or eight he would tell me exactly what he wanted me to wear. He had dresses he didn't want me to wear at all and others that he loved. If I'd put the wrong one on he would kind of shake it, meaning "no, no." Then I would say, "All right, what do you want?" and he would go to my wardrobe and pick up another dress and give it to me to wear. I would wear it just to please him. He liked dresses that had a lot of red in them and that were very bright.

By the time they get to be six or seven they get much more crafty than you are. I would lock all the cupboards because the maids would steal. Tuan would get the keys and he always found the right key for the right lock. I don't know how he did it. He had to operate fast, and he knew in what particular cupboard there was a biscuit tin, even though I would change its location. He would take the tins out, open them, stuff his mouth full, put everything back, close the cupboards, even lock them, and hide the keys somewhere. We had to spend hours looking for the keys. Of course, there would be the telltale crumbs all over, and I knew it was him. I would scold him and he would look at me with "Oh my God, how did she know? How can mommies be so clever?"

When we left Indonesia, the orangs could not be taken out of the country. It was terribly hard on them because they had always been humanized and suddenly they found themselves being treated like animals in a cage. I visited them at the zoo two years later and they went crazy. I realized it was very bad to visit them. Dulu in fact died about one month after I left. I think he just gave up food when he saw me really go. They were in such a mess I couldn't stand it. You could see all the ribs on Tuan, and he looked like he had skin disease because he would roll in the cage and lose all his beautiful hair.

After leaving Indonesia, my husband and I landed in Singapore before going to Europe. My first visit, as always, was to the zoo. There I saw fourteen orangs in a very small cage, about 12 feet square. They were all very humanized, and just a fantastic, highly intelligent group. I stayed in Singapore for seven months and went to the zoo every day from early morning to late at night. I decided to try to understand their communication and behavior better. The zoo officials allowed me to enter their cage to do my studies. 59

When the dominant female came to puberty she was bleeding, and she immediately thought she was sick. The first thing she did was refuse to eat and drink, as they do when they are sick. Then she kept showing me the blood. She would have it on her fingers and put her fingers under my nose, and I would say, "Yes, I know. I know you're bleeding. It's all right. There's nothing wrong with you." But she was so upset, she kept knowing there was something wrong with her. All day she kept showing me the blood. "You see, I am still bleeding. I am still bleeding." Her expression was just that of anxiety. By the end of the evening she believed she was going to die any minute. Finally we said, "We have to do something. There's nothing wrong, but she thinks she is really sick." We decided to use psychology. I brought the doctor's little black bag and gave her a vitamin injection. After that she immediately started eating her usual meal. She knew now that I was looking after her and that finally she was being treated for what was wrong with her. The wonderful thing about them is, they remain childlike always. A fully grown one, capable of raising its own child, is a child herself. They have that beautiful innocence that certain human children have. I admired most that dominant female in the group. She was the most intelligent, fair, sensitive, and alert. In fact, all the zoo people liked her so much that when she died at three in the morning while giving birth, they were all crying. None had gone home.

From Singapore we went to Switzerland, where my husband continued his work, but our marriage was not working out. There I had only little guinea pigs in our apartment. I had been lecturing on primatology and decided to come to America and start a sign language program with an orang. By then my marriage had ended.

It took almost a year to start a nonprofit organization, to obtain the special permits for orangs, and to get the Yerkes Institute to donate an animal to me. Orangutans are all individuals, and you never know what type you are going to get: a selfish one or a generous one, a dumb one or a bright one, a tough one or a sensitive one. But I knew I just couldn't live without one. They gave me Bulan, who was one and a half years old. "Bulan" means "moon," an Indonesian name he was given at the institute. He was born at the Yerkes, and he knew his mother for only three days because she chewed his umbilical cord so much that she chewed a big hole in his abdomen. He had a skin graft that didn't take, but he survived. Then he was caged with another orang that was used for biological research. He was very antisocial when I met him, and very, very afraid. My assistant went to the airport, and when he opened the cage, Bulan just jumped on my lap and stayed there and hung on and hung on. He had not had a mother, so he missed that terribly. And he decided I would be the one. Initially, I carried him everywhere I went and slept with him at night. When he saw people he would get

himself all dirty, and I would have to receive them full of you-know-what all over me. I couldn't even go to the bathroom and change because he would hang on so. It took me about half a night just to undress to go to bed. He was tremendously insecure and afraid to lose me. Now that he had a mother, he was going to hold on to her regardless of what happened. Unfortunately, he is of mediocre intelligence. It is a pity, because people will judge orangs by him in many ways rather than by a brighter one.

A typical day begins at 5:30 or 6:00 A.M. He climbs into my bed and tries hard to wake me up even if I pretend I am asleep. He will undo the covers, shake me, and make quick calling sounds. I get up around seven and he will go straight to the toilet. Even though in the wild they have no fixed spot, he will squat on all fours on the toilet. Then it is straight to his bottle. He drinks milk and egg yolk. By nine he wants his breakfast, which means solid food of some kind, either bread, beans, or vegetables. He is very impatient. As soon as he sees the food he signs, "Quick, quick, hurry, hurry, hurry." "Hurry" is one of the first words he learned. He will look around the stove and see what is cooking. He will sample one spoonful and decide he doesn't like it and spit it out all over the floor, then steal something sweet from the cupboard and run away with it. Then by eleven he is getting his milkshake; that is when I mix in all the foods he doesn't normally like.

Then there is his sign language lesson. I began with him when I got him. You do it in three stages. First you pick up an object like a glass and show the sign for glass. Or you tickle him and show him the sign for tickle. You don't expect him to do anything. The second stage goes on for about two or three months. You start molding his own hands to make the signs. You show him the sign, then you mold the hands. You do that time and time again and then eventually, after many repetitions, one day you will be rewarded with a sign. His first sign was "drink." Holding the thumb to the mouth with closed fist. His second sign was "love hug," which is typical of him because that is what he needs most. It is a crossing of both arms over the chest as if you are hugging yourself. He has nineteen signs, some of which he hardly uses. He uses "hurry" and "love hug." "Drink" he uses only if asked. He can sign "Bulan," which is the right hand on the shoulder. He calls me "you," meaning that he points with his index finger.

One problem is that they don't see the point in some of the ideas. Why count up to five, for example. They think "five" oranges. Who wants to eat five oranges anyway? Three is more than enough. In the wild they have about forty sounds, but they use only ten or fifteen on a regular basis. Bulan makes some sounds on his own. The loudest sound they make is the territorial mating call. It can be heard a mile away. They inflate a sack in their chest and then rhythmically shake the breathing components. Bulan is practicing with this a bit now. It is a gentle sound so far, but it is getting louder. I watch 61 him from a distance, because if I come close he stops. By 6:00 P.M. he becomes hungry and thirsty and will take my hand and lead me to make food. Often I am gardening, and he will cry and cry.

When he was three, he needed a lot of discipline. I was far too lenient with him. I tried several ways and it didn't work. Then one day I realized he was afraid of water because in Yerkes they hosed the cages and it terrified him. So, I thought, "Well, why don't I punish him with water?" I poured water on him and he was extremely surprised. Then he realized I was angry. The second time I threatened him with water he disappeared and came back with an umbrella. Then he played with the umbrella and pulled it apart. The next day there was no more umbrella and I threw some water on him. This time he looked around the house and found a huge plastic sheet and covered himself with that when I threatened him. He does show tool-using. He knows how to start the car engine by putting the key in, turning it, and pushing the pedal. He also knows how to take off the little cap on tires and push the air out. This is very awkward because you never know if you've got a flat or if it is just Bulan.

They investigate themselves when they are young and they masturbate, not so much for sex as for playing around with the different sensations of different parts of their body. They have the advantage of having two feet that are also two hands. They can be sucking a bottle and doing that, cuddling "Mommy" and doing that, or watching a movie and doing that. I don't stop him because I know he is not going to do it in front of other people. But strangely enough, he does do it in front of a certain lady. He wants to kiss her on the lips and hug her all the time, and he gets a sentimental look on his face. It began when he was two and a half years old. She is very proud of it. She says, "My goodness, to be loved by an ape." She was very beautiful when she was younger, but now she is not so beautiful. I don't know what it is about her, but it's happened since the first day he met her, and every time he sees her he starts all over again. Another thing he will do with women is look under their skirts, but that is curiosity and has nothing to do with sex.

I am hoping someday to get islands where I can really contribute to the orangs' preservation. It would be so easy to keep them. They don't need a lot of space, just a wild habitat surrounded by water moats so they won't escape. I want to make the public aware of the danger of extinction and that we have to do something. The great apes are our closest relatives. We have to preserve them to try and understand ourselves, our prespeech and our presocializing. Maybe the orang is further away than chimps; research seems to point to that fact. But somehow I prefer to see human beings more like orangs, with the more noble characteristics the orang has.

62   My greatest moments are when I am sitting beside Bulan trying to teach

him something, and he doesn't really want to learn it, and he is making little squeaks that say, "Don't bother me, let's just cuddle and love." He looks at me with a sentimental face, a really adoring face. Anybody coming into the room will spoil everything immediately. His mood will change and he will become regular. Those moments are like a honeymoon, when you are with your husband alone and lying in bed and you look at each other and you know you will just adore each other for eternity. It is that same kind of extremely close relationship.

The Smallest Horses in the World

At Hobby Horse Farm, Robert Pauley, forty-one, and his wife, Jean, thirty-nine, are carrying out their dream of breeding the smallest horses in the world. Their children, Robert Jr., fifteen, Tammy, eleven, and Mark, six, have had the horses as playmates all their lives.

The Pauleys' 435-acre breeding farm in southwestern Virginia looks out over the Blue Ridge Mountains. Their home, a red brick, three-bedroom, ranch-style house, is cozy, with Early American furnishings. Over the fireplace are many trophies from the International Miniature Horse Association: Grand Champion Stallion in 1979, Grand Champion Mare in 1979, and Grand Champion Stallion in 1980. Bunches of blue ribbons are also displayed. The tiny, fancy saddles and bridles used by the miniatures hang over the arms of the easy chairs in the living room.

ROBERT PAULEY: I always liked small animals — squirrels, rabbits, anything. I remember that I asked for a very small pig when I was probably seven. My father brought one home in a coffee sack. I made a wish that it would never grow any bigger. Believe it or not, it must have been a runt, because in six months it grew very little. Then all of a sudden it grew and made a 350-pound hog and had to be killed for meat. But I never did eat any of it. Later I had rabbits, some bantam chickens, dogs, and a jet black quarter horse. Somebody picked him out for me and my brother, who was two and a half years older. He was the blackest horse I've looked at — velvet black with a white star. The best I can remember, we called him Black Beauty.

Miniature horses have been around from the late sixteen and early seventeen hundreds. They were first bred in Europe and were very rare. Only kings and queens and members of royal families had them as pets. Later they were used in circuses. They've only become popular with people in general in the last few years.

In the early forties there was a gentleman, Mr. McCoy, in West Virginia, who was breeding them. He put a $1000 reward in a magazine for anybody 63 that could prove him wrong that he had the world's smallest horses. No one won it. Then another breeder in Bedford, Virginia, Mr. Field, developed some. Mr. Field and Mr. McCoy were the two pioneers who started miniature horse breeding in the United States as far as I'm concerned. They had brought their original stock from Holland.

My wife and I heard about miniature horses many years ago, when President Kennedy's family got a couple. After about a year and a half, I finally found three at one place and two at another. Then I got more enthusiastic and bought out another breeder and got twenty-one more.

I wasn't originally a farmer. I owned and operated a body shop and lived on a farm near Richmond. I didn't particularly want to make miniature horses a business because I had the body shop going real good and I had six employees. But it got to be publicized that I was breeding miniature animals, and the sales just started. I didn't end up missing my business because I was always looking forward to coming home in the evening and playing with the toy horses. My wife and three children like them, too.

We're going on sixteen years now. As we've reduced their size, we get more and more attention. The television people came out and did different articles and stories, and also the newspapers and different magazines. They have been in several publications including the *Globe Magazine, National Enquirer*, and a German magazine. This helps to promote the breed and let other people learn about it, and we've sold horses all over the world as well as all over the United States.

They start at $1000, then go up. The smaller they are the more expensive they are, because when you get down in size you're looking at a smaller number of horses. A very small one could sell for about $15,000 and up. I wouldn't sell our miniature Arabian-type stallion, but his value would be somewhere near $100,000. We're getting these perfect Arabian offspring out of him that will mature under thirty inches.

As pets they like attention, and the more time you spend with them the gentler they get. Some follow you around like a dog and come when you call their name. In the summer our little Cricket will come in the yard. If the children have a basketball, she will get on top of it and scratch her stomach. When she goes back on the basketball her front feet come off the ground, and when she goes forward her back feet come off the ground. She's got really short legs.

They don't need shoes, but if some people drive them in a parade or on the pavement they glue on rubber ones. You just file their hoofs down like fingernails. And you can feed about six miniature horses as cheap as you can feed one regular riding horse. You have to get the harness special made, and usually a good set runs about $400. Some people make little wagons and miniature stagecoaches for them. We have sold them at Christmas for people

who actually put them under the Christmas tree. They live an average life of twenty-five years. The oldest I've heard of was forty-one.

You don't have to hand-breed them like regular horses, where they have to worry about the mare injuring the stallion, because with a miniature horse you've got a miniature kick and no one can really get injured. Though once I had this little bitty stallion that fell off, but I guess the only precaution is to catch them if they fall.

If you breed too close, you wind up with animals that are dwarfy. This has happened to me only once. I ended up with a little horse with an oddly shaped head, short neck, and short legs. But she was kind of cute. When she was born she looked like a stuffed teddy bear with 8-inch legs. A lot of people wanted to buy her and give her a good home. So dwarfs have a purpose, too, but that's not what we are breeding for. My feeling is, if an animal has the slightest chance of making it and not suffering, I'll keep it alive. She got special care because she was so small, and she was very gentle and tame. But you never breed a dwarf.

We have about twenty real good little stallions that are 26 to 28 inches, and we have enough different bloodlines that we don't have to worry about interbreeding for the next fifty years. And we can trace the pedigrees back to the early forties. We also breed for different distinct types: Arabian, Belgian, Appaloosa, quarter horse.

My favorite is Toy Boy. To me he is like a perfect five-carat diamond. No flaws. He's priceless. He is elegant and fine-boned with a small head and dish face. Across his nose is approximately 3 inches. And he's a 27½-inch horse. His hoofs are about silver dollar size at the maximum. And I'm very happy about his son, Golden Toy.

Tiny Tina is one of the littlest horses ever born. Right after she was born and first stood up she was about twelve and a half inches. But after she stood up good she was fourteen inches. Now that she is grown she weighs about 55 pounds. Her mother wasn't so small, she was 32 inches tall, but colored pinto like Tina. I've had one man that owns miniature horses, that has been down here about a dozen times looking at her and still can't believe it. They put a picture of her in the *National Enquirer*.

There is a registry for miniature horses called the International Miniature Horse Association. I helped support it from the start and have been a director for several years. The standard is that they have to be in proportion to a large horse; they cannot be freaks or dwarfs. They cannot exceed 34 inches in height. I would say there may be twenty-five hundred horses 34 inches and under in the country. Then, as you reduce that size, you reduce the number tremendously. You go down to 32 inches, then you're looking at maybe a thousand horses at the most. When you go down to 30 inches, then you're talking about two hundred and fifty to three hundred. When you get down to 28 inches, you can start counting them on your fingers. There's just not that many.

Each herd has a stallion. Several years ago I remember June Bug, who could tell how many mares he had. He would bring them all in in the evening and he stood and counted them. One time I saw him scanning the pasture and he knew something was wrong. Now they are so short that if they put their heads down like a goose they can't be seen over the high grass. But he rounded up that last mare; it was like he knew, "I've got thirty-one, and one's missing." Maybe he wasn't counting, just looking for the different faces. They're smarter than you think. For instance, if you have several mares out in a pasture, if one is ready to go into labor, there is always another mare who pals up with her; we call it the "protection mare." She helps kick at the rest of the herd to keep them away because they are inquisitive when a new one is born. The mother is too weak to do this. Later the mother and the protecting mare shelter the baby in between them. A couple of days later the protecting mare goes off and they're not buddies anymore.

I don't think anybody should have to grow up completely. If you're sixty years old, you should be somewhat of a child at heart. Probably there was enough child still left in me that the miniature horses really fill that wanting of small, fairy-tale animals. I got my brother interested in the small horses, and now he's got about twenty and just loves them. They are like a very small breed of dog: There'll always be a big demand for them, and you just can't raise enough.

I would really be pleased if I could get a nice small herd that was, say, 20 to 24 inches in height. I think that would really be doing something. We have perfect horses from 26 to 28 inches now. To drop to 20 would take another ten to forty years. But I am satisfied to know that I am making a contribution to the smallest horse breed in the world. And best of all, it's always like Christmas - each spring when the mares are ready to foal you don't know what surprises are waiting for you.

Can One Measure Happiness in a Frog?

Amphibians are dependent on both aquatic and terrestrial environments at different stages of their lives. They hatch from jelly-coated eggs in ponds, streams, and lakes, and as babies they are fish-shaped and gilled. As they grow, hormonal changes cause legs and lungs to form, and with adulthood they venture onto land. But because of their delicate, glandular skin, all but a few species require damp and sheltered environments.

In two 12-by-16-foot basement rooms of his home in Columbus, Ohio, Frank Juodvalkis, forty, has more than twenty-five meticulously maintained aquariums ranging from 2 quarts to 400 gallons. Each is a perfect miniature world, with an artful selection of precisely arranged rocks and plants. Fluorescent and Gro-Lux lights are used for the plants, and incandescent bulbs provide heat. He keeps the ambient temperature at about 65 degrees in the winter and in the summer about 10 degrees cooler than outdoors. In the hot season, the wooden heat-conserving tank covers are changed to screen-type tops.

Frank maintains records on the feeding patterns of each animal. Their species include: seven European green toads, two Running toads, six Asian red-bellied toads, three Turkish tree frogs, two Siamese painted frogs, four Spring Peeper frogs, two Surinam toads, two Congo eels (amphiumas), two giant sirens, three dwarf sirens, one yellow-stripe Caecilian, six Eastern spotted newts, three European crested newts, four Japanese red-bellied newts, seven Spanish newts, three Oregon newts, three tiger salamanders, three spotted salamanders, five Burmese salamanders, three Midwife toads, two Cuban tree frogs, three mud puppies, and a 22-inch hellbender. He also has some unusual fish. Among them are eels from Asia, North, and Central America and a breeding pair of South American lungfish, which are almost 2 feet long. The turtles he keeps are aquatic, as are the rare freshwater Southeastern Asian tentacle snakes and the elephant-trunk snake.

Each week the animals eat over two hundred night crawlers, which Frank catches himself. In addition, he raises mealworms and gets crickets from a friend to supplement the diets. He also makes a paste of beef heart, frozen smelts, gelatin, vitamins, and minerals, which is frozen in a flat sheet and fed in chunks.

Frank and his wife Teresa, thirty-nine, have been married for eighteen years and have three sons. Family pets are Bruce the cat, Feather the cockatiel, and Otto the dwarf hamster.

FRANK JUODVALKIS: I can't tell why amphibians and salamanders in particular appeal to me more than a snake or a dog or a bird. I think it's their primitiveness. They echo something very old, and the appeal is not even on a conscious level.

Amphibians obviously are not the most colorful animals, yet I am fascinated by them. They usually move very deliberately and slowly. They are hiders and burrowers. It is very unusual to see a salamander out in the open. To look for them, you lift rocks and roll logs. That has always appealed to me. It reminds me of a little boy mucking about in the woods. At times I go out and collect animals — not to bring them home, just to look at them. I don't know if I can put my feelings into words, but it is like when a little tree frog holds on to your finger, it's an almost paternal feeling. I also sense it when those trusting, beady little black eyes of a salamander stare up at me. I feel an affection for the animal. I am amazed that amphibians as a whole are still alive today because they really have so few defenses and they seem to be predated upon so readily. Of course, the toad has parotid glands 69 that have a protective bitter secretion, but the salamanders and most of the amphibians that go through the larval stage are food for every fish in the pond.

An important part about my keeping amphibians is that they don't require daily care. I am a salesman selling industrial tools. I am on the road all day, but normally I am home every evening. Probably on the average of twice a month I will be gone for two or three days, and those are mostly for things like educational seminars or trade shows. It won't hurt my animals to go a week or more without food.

In Cleveland when I grew up I didn't even go to the zoo. I was a city kid, a typical street kid. We didn't live hand to mouth, but we didn't have money for things like aquariums even if I had been interested in them, which I wasn't. My mother raised me and my brother, four years older. We are immigrants. I was born in Lithuania. We came to the States when I was almost seven. My mother worked from eight to five at a bank in Cleveland. My brother and I went to school, and then out of necessity we were on our own a lot. My mother was a widow. Mom remarried after I got married. My father was declared legally dead 'cause they never found the body. He was in a bombing in World War II. People knew he was in a building before this particular air raid, but nobody ever found the body. It was assumed that it was lost under the rubble somewhere. I've seen pictures of him. He was a very handsome, dark, tall man and he was very athletic. He was much better looking than I am.

My wife was the first one who brought animals into the house. Somebody gave her some guppies about fifteen years ago. I stayed up till dawn one night watching guppies being born. That really tickled me. I had never seen anything like that before.

After getting started with tropical fish, I started reading about amphibians. The last chapter of aquarium books would often mention them. When we moved into our present house I gradually expanded my aquarium setups, but with amphibians instead of fish.

One of the advantages of liking weird animals is that you get a reputation for liking weird animals. For example, when a tropical fish wholesaler would get five or six unusual salamanders, he would call me and I would get them all. The same thing happened with a couple of giant sirens. A guy bought "lung fish" from a fish dealer in Florida, and he knew enough to know they weren't lung fish, but he had no idea what they were. So, he called me and I went down and said, "Hey, you've got giant sirens." We made a deal right on the spot. A nice thing, too, is that the specialty isn't crowded. But it would be nice if there were a few more people in Columbus who had the same type of interest and we had a pool to trade animals, collect information, and disseminate it.

I'm familiar with most local amphibians. If I'm looking for a tiger sala-mander, I know when and where to go. For example, I go for the larger salamanders in the spring when most of them are breeding. A good thing about living in Ohio is that there are so many species of salamanders. Gen-erally, they breed in temporary ponds. The tiger salamanders breed in deeper ponds than most of the others, but usually where there are no fish, so there are no predators on the larvae. To collect some, you go right after the snow starts melting, especially if there's a thunderstorm in late February or March and there's a lot of water. You walk on cracking ice in shallow water at night with a flashlight and net. My problem is, I've got big feet and I can't find hip boots my size. Years ago I should have made the investment and had some specially made. It's really a form of masochism the way I do it.

I prefer a terrarium-type setup with water, land, and plants. With am-phibians, you want to keep them relatively damp and relatively cool, de-pending on where they are from, of course. I like to set up a swamp or bog environment. It is not the easiest setup, but you get a feel for what will make the animal comfortable and what won't. For example, for any of the larger aquatic salamanders like the amphibians, sirens, hellbenders, and mud pup-pies, it is very important that they have a roof over their heads. If you have an aquarium with just a rock ledge, they will stay under the rock most of the time. But if you put in some floating aquarium plants with the rocks, they will come out from under that rock very often and will feed readily.

In some ways amphibians are like the kids; you don't realize how fast they grow. What I do now is I keep records. I measure animals as I get them, and periodically I make another entry in my book so I can see over a period of years what kind of progress they are making. I also keep a schedule. I chart when I change the water, when I feed, what I feed, what each one eats or doesn't eat. I can tell everything at a glance. If something goes off its feed but it did the same at this time last year, you don't get quite so excited, or if it is atypical for something not to eat, you know you should be concerned.

Usually I go down in the basement when the kids go to bed, my wife is reading or something, and the house is getting quiet. It's the way I wind down. I go in and turn on the radio and putter around for a while, like some guys do with electric trains or water colors.

Amphibians do well for you if you are willing to put in the time. I keep them relatively cool. My basement, even in the summertime, is 10 degrees cooler than outside ambient, which is sufficient for most of the amphibians. I feed everything by hand individually. Now, if you have a tank full of fish, you can throw in some food and watch them eat. You can't do it with little salamanders. I try to prevent them from becoming obese. Some, like the Japanese fire belly, will just bloat themselves and eat as long as I keep feeding them. The time I spend in the basement varies because it is rare that I will just go down there and feed. I will start and then say, "Hey, something needs to be done here." So I will stop feeding and change the water or change rocks. And I don't try to feed them everything all at once. I will go down and feed smelt to the fish, amphiumas, and tiger salamanders one day, and the next day I will chop night crawlers. The following day I will throw in mealworms and crickets and so on. It is a varied cycle. I like it that way.

The difficult part of breeding amphibians is that they have so many offspring. They can lay up to several thousand eggs at a time, and I would feel obliged to try to save them all, and that's an awful lot of work. That's why it is easier on me if they don't breed. In the wild, only a very small percentage survive. A bunch of little toads hatch out and hop around, and every hognose snake in the neighborhood eats well! But I feel bad if my animals die, and rather than harass myself, I just try to maintain them healthy and looking good so I can enjoy them. I don't keep pets because I believe it is just for their own good. It is for my own good, too.

Hellbenders are my single favorite animal. I have a friend, affectionately referred to as Tommy Turtle, who goes out and knows where everything wild can be found, whether it is a fish, salamander, or a turtle. Many years ago I told him I would love to get a hellbender. I had never had one. So we went out in October into cold, cold water, waist deep in a trout stream. The water flows fast and you have to walk carefully. You bend over large rocks and reach under, trying to blindly feel around for something soft and squeechy. It is either a carp that has been dead for a week or maybe a hellbender. If you think it is a hellbender, you try to find where the head is 'cause you don't want to get bit. They have teeth that are very sharp. You grab it by the underarms and you pull it out wiggling hard. With luck, your friend is there with a bucket and you get yourself a hellbender.

I had one hellbender for five years; he died five or six weeks ago. He looked so ugly and repulsive that he was appealing. He was 15 to 18 inches long. Somebody likened him to a rotting cucumber, and I think that is appropriate enough. Yet in a bizarre kind of way he was attractive because his loose folds and flaps of skin were very wrinkled. He was kind of clumsy, a creature of the black lagoon, and just so old-looking that you could almost see the moss and hear the crickets when you looked at him.

This hellbender stopped eating, which he had done occasionally before, and I really wasn't all that concerned. But then he started getting thinner and deteriorated so quickly that I wasn't surprised when he died. It could have been caused by some kind of internal parasite, disease, old age, or even cancer. It was something over which I had no control. The bad thing about having an animal that long is that it's like losing a dog you have had for five years. You really miss it. I don't get embarrassed telling people that I am really sorry I lost this animal. I know people with, say, snakes or birds are embarrassed to show that they miss one. They wouldn't be embarrassed if it were a dog or a cat. I knew I wasn't irrational. I was upset because it was my single favorite salamander. I didn't go through a burial ceremony—I just cried as I dropped it in the garbage can.

The animal I have kept longest is a female lung fish that I have had for fourteen years. My oldest boy is thirteen, and I tell him everything works here on a seniority basis, so he knows where he rates! The lung fish was very small when my wife bought it for me as a birthday present. I think animals in captivity tend to live longer because of no predation, no accidents, and if they are kept well, no diseases. They live healthier and longer and get bigger, but I don't know if that means happier. Can you measure happiness in a frog? But you do get a sixth sense and can tell if they are content.

I gauge intelligence in amphibians by how much they will react. The celicans have very little reaction. They just hide. But I have had aquatic celicans that did learn when it was feeding time, and they would come out of their burrows. I think ambystomas are the smartest of the terrestrial salamanders.

I can't think of any Ohio amphibian that is endangered because of collecting and none I can think of has any commercial value. But when most people ask me where they can get a hellbender or a tiger salamander or something like that, I don't tell. There is enough pressure now on populations primarily through habitat destruction. Sometimes I wonder if I am doing amphibians a service by talking to people about them, by trying to kindle an interest. I hope that any danger is offset by an appreciation for them, but I think the worst thing that could happen is somebody getting turned on and going to rape the woods of everything that moves.

It is hard to explain why it would be a loss if amphibians were extinct because people as a rule put things in economic terms. If they can't eat it or wear it, it is not good for anything. I don't quite follow along with that. I don't know that I would be any better off financially if there were no tiger salamanders in the whole world, but it wouldn't be the same kind of world. I don't want to sound melodramatic, but it's as if you loved music and all of a sudden there was no music. For me, something would be lacking even if I couldn't quite tell what. My message is, there is no such thing as a worthless animal.

A Lifetime of Circus Bears
and Rare White Tigers

Herta and John Cuneo are the third owners of Graze Lake House in northern Illinois, where they have lived for six years. Set far back from the road, the fieldstone and glass house is one of Frank Lloyd Wright's later homes, which he designed in conjunction with his son, Lloyd Wright. The entire property encompasses 260 acres, and the 75 acres around the house are fenced to enclose herds of fifteen pure white elk, thirty all-white fallow deer, eight axis deer, and three buffalos, two females and one male.

For all of her forty-five years Herta Cuneo has been a circus person. And her husband of thirteen years has been a circus fan for most of his life. Now, in addition to many other business ventures, he leases their elephants and tigers (both yellow and white) to circuses. The Cuneo animals appear in the Circus Vargus, the Shrine Circuses all over the United States, and in the Hamid Morton Circus.

Inside, the Cuneo house is opulently furnished. Original works of animal art line the walls. Herta's dogs, Fritzie the dachshund and Heidi the black and tan Doberman, follow her everywhere. On the top floor is Herta's nursery and office. There, in a green-carpeted room with yellow-and-white-cushioned furniture, she mothers the baby tigers. When I was there, the nursery was occupied by a five-week-old litter of three, two white with blue eyes and one yellow with golden eyes. They have their own small playpen, and Herta bottle-feeds them several times a day. When the tigers are older, she will swim with them in the huge outdoor pool.

Twenty-four miles from the house is the 25-acre animal compound. There the Cuneos constructed a large warehouse that they divided in half. One side is used for elephant training, the other for tiger training. The compound also contains separate buildings for the tigers, elephants, and bears.

Herta no longer performs, but a few times a year she visits the circuses in which Cuneo animals are performing.

HERTA CUNEO: I was born in the circus to circus people. When he was just a kid, my father actually ran away to join the circus. He was Austrian, but worked for a German circus. His mother didn't think much of circus people, and she brought him back home, where he apprenticed as a sheet metal worker. But the minute he got his diploma he left and rejoined that circus.

He liked the animals, so he started out being a groom for tigers and lions. The circus had a bear whose owner was old and was looking for a helper; my father was probably twenty-three when he began to work with him. My mother's family owned a Munich beer garden where he would eat because they let apprentice trainers get free food. That is where he met my mother. I don't think she had an interest in circus animals, but she had an interest in my father! When they married, she got to like them. In fact, she told me that she worked in the show till a month before I was born, and three days before I was born she was still rehearsing with the bears. I don't have any brothers or sisters. In those days the circus owned the animals, and they hired the people to train and present them. So my father worked there for sixteen years but the bears were not his. When I was about seven he left and bought his own bears and moved back to Austria. Our family performed in Hungary, the Middle East, Egypt, Syria, and Lebanon. Then in 1949 they joined Ringling Brothers. My father was the first to train a bear to ride a motorcycle.

In those days Ringling Brothers still had a big top and traveled in four trains: one for the performers, one for the animals, and two for the tents and equipment. The animals always left first and the performers last, and the acts were only one-day stands. That meant that your animals left immediately after the act, and you didn't see them until maybe an hour before showtime. My parents quit after the first year because, coming from Europe, they were used to the animals being their whole life.

My mother made costumes for me and for the bears. I liked spending my time with the circus animals. When I was about seven I was allowed to go on the ponies, and the elephant trainer would let me go for a ride on his animals. We children would imitate the acts from the circus, and the adults would teach all of us tumbling, acrobatics, juggling, and contortion. I began working in my parents' act when I was eight. My father would have a bear walking on his hind legs on a ball, and I would walk on the ball too. I was never really frightened of the bears. I was more frightened of my father because he would get mad if I got hurt. He would always say it was my fault, which it usually was.

At sixteen, I actually started helping him train the animals, but he never thought I was good enough. Then he got very ill with several heart attacks, and I had to continue the act while he was in the hospital. He would always say the bears weren't working as good as when he left. In those days I was like an extension of him because although he wasn't physically able, he would sit there and tell me what to do. After he died, when I was thirty, I realized how much effort it really was and how you hate to see somebody else working your animal and possibly messing it up. Then I understood him much better.

About the time my father died, I married my husband, John. His father had many businesses and even was distantly related to the man who started

Bank of America. The family had a lot to do with animals because they owned

Melody Dairy in Chicago and bred horses. We have never had children, but my husband has a son by another marriage.

John, who for many years was a circus fan, and I met in 1950 at the Chicago Railroad Fair; he was nineteen and I was thirteen. I was performing with the bears and he was watching and talking to my father. First I wasn't allowed to talk to him at all, but he told my father that he had two American black bears at home and my father said he would like to see them. He took us to his home in Libertyville and, sure enough, he had two bears; one was trained to ride around in a little jeep. I was the only one in the family who spoke much English, so I acted as translator.

He married twice, and I knew both of his wives. After they were divorced, his first wife got killed by an elephant, right in the ring during a performance. She really humanized her animals completely and would work with five male elephants. There are very few people anywhere who can handle them well, and it had gotten to the point that she had to sleep on a bale of hay in front of the elephants because if she left them, she was terrified that some person would get killed by them. They were that evil. One of them killed her up in Canada. He totally skewered her — killed her on purpose. He knocked her down and rammed the tusk into her chest. They say she died instantly. They had to shoot the elephant to get her off because he was just standing in the center, twirling her around and bellowing. She always had a thousand and one excuses for the animals, but they didn't respect her as boss.

When my father retired, I was twenty-nine and had never been out on my own. I talked to John, who said, "Why don't you come work for me? You can work the bear act, and can work with me with the dressage horses." In circus life, most things happen during the winter because during the summer you are out performing. I thought I would have time to learn to ride. But as it happened, when we combined the acts he had a show scheduled, and his regular girl was about to have a baby. That left me five weeks to learn the three-step, Spanish walk, pirouette, trot, and canter on well-trained Andalusian horses named Turko and Prince. Now, when you train a bear there is not constant physical contact, but when you are sitting on a horse you have to tell it what to do all the time. I had one week on a quiet, old, retired dressage horse that had to endure me learning the signals; then I got on Prince, and it was like moving from a tired old pick-up truck into a race car. Well, after five weeks I was so sore I could barely walk, and I thought that horses were the most stupid creatures compared to bears. Then we had the show, and people came back afterward and said how pretty I looked. I had on a long gown with a top hat and a veil, and Prince was a beautiful horse. They would say, "How long have you been riding?" And I would say, "Oh, five weeks," and they would think I was putting them on. 77

I have had different kinds of bears in my act. The Syrian is a popular kind to work. They look good and are quite smart. Russian bears are popular, and I had one of those as well as grizzly and Himalayan. The Himalayans were never very safe. I don't think bears were put on earth to live that closely with human beings. Yet I really am against the way some animals for circuses or movies are declawed and defanged. They should call it what it is: deforming an animal. If the claws of bears are taken away, they don't have the same balance, and if their fangs are removed, they can't tear their food properly and can't work up the saliva that helps their digestion.

I believe that in training bears there has to be a reward and punishment system. With my bears I used two 3-foot iron rods, a bigger one with a blunt point to use as a prod if they charged, and another one to hit them on the front feet or the nose. Normally, you rarely need to do anything, because you yell when you discipline and they learn to respond to the tone of your voice. For rewards I used M & M's chocolates, mint candy, or cookies.

I used a cattle prod on a bear only once, and that was one of the times I almost got killed. I was training little Teddy to ride a bicycle. It is one of the dullest tricks you can teach because it goes on forever. First they learn to walk on their hind legs and sit on a little chair, then on a saddle. Then you have to put them on a stand, so the back wheel is free, to teach them to pedal. To get them to keep pedaling you move their feet, then let go a little, then give them a goodie. Eventually they get it, and you have to teach them to balance while they pedal. Teddy and I were doing this for eight or nine months — a long, long time. Finally she did a figure eight in both directions, but she would go fine for one or two rounds, then get slower and slower and stop pedaling. That is when I thought, "Maybe I am going to try a cattle prod." The normal way to do it is to run behind them and push on the bottom of the hips with your fingers. But every time I would quit, she would quit. Well, I zapped her with the cattle prod, and she dove straight off the top of the bicycle — just raged up in the air — and came flying at me. She had a leather ring muzzle, which they have when they perform, but they can still open their mouths. I got some scars from that time, and I have never used a prod again.

Bears are not demonstrative and are not very affectionate. But I have liked all my bears, especially Junior and Whinny, which I raised, and their little mother, Teddy. Their father was Boobo, who died last year. He was a very big Syrian male. I now have seven, all the surviving bears I worked with. I feel like they have been members of the family and have made a living for me, and I owe them something. There are no ways of placing old animals. Most zoos don't want bears because they are not that rare, and these can't be put all together in a little group like zoos do. They live to forty or fifty. When they retired, we built each one a big cage with an outdoor swimming pool.

Even before I joined him, John said we should try making a tiger act. I really wasn't that interested since my father had been unlucky with them. You see, there is a totally different system in training them than with bears, because you have to keep a much bigger distance. With bears you have to physically make them do things, like if you want them to do a headstand you have to lift the rear end up. That is why you have to start when they are young. And this might sound strange, but when they hit you, it's better if you are close, because they don't have retractable claws. They slap, and if you are closer you get the paw part, but if you are farther away you get the claw end and get a really bad bruise. Later, when I began to train tigers, I wanted one to roll over and she didn't want to. Instead of getting farther away I kept on getting closer and closer, like I would with a bear, where you would physically take him and turn him. When I got too close, she leaped up and just split my whole hand with her paw.

When our tigers started having babies, the first they had was white! We didn't try for it. The first white tigers were discovered around 1953 by the maharajah of Rewa, India, who caught one, bred it to yellow tigers, and eventually got a lot of white ones, some of which were given as state gifts. We had gone to visit the Washington zoo and the two things I wanted to see were the white tigers and the panda bears, which they had just gotten from China. Then I flew to Florida, where I got a phone call from my husband. He reported, "Sheba had babies, and they are white." My answer was: "Oh yes, and Teddy and Boobo had babies, and they are panda bears!" I wouldn't believe it until I got there. One, unfortunately, died of a heart deformity. We had no more white babies until four years later, when Sheba had Frosty. We also bought Sheba's sister's son, Tony, and have bred him and gotten white babies.

We can't be sure that ours are related to the ones from India; there may have been a different mutation that made the color. A study is presently being done on white tigers. Most of them become cross-eyed as they get older. Frosty is badly cross-eyed, but Silver, who was born last year, is still good. They also look bigger because they have longer legs.

We now have fifty-two tigers. Even our yellow babies now have white genes, so we hope to produce more white tigers. My favorite activity now is raising all the little tiger kittens. I take them from the mothers at birth because we have had mothers kill babies up to two and a half months old. I bottle-feed them and keep them in the house. But once they start eating meat, they smell pretty bad. They also start tearing up the furniture. When they are older, they swim in the pool with me. At four months their training starts. It is unpleasant to me when my babies leave home to go to the circus.

Our circuses pay for themselves, or at least enough for the feed. It takes between 15 and 20 pounds of meat a day for the tigers, and that is quite expensive. We buy meat at 6000 pounds a time, and our trucks have huge

freezers. The present problem is finding trainers. It's no longer like the European tradition of the children staying in the circus. But without the animal acts I feel something special would be lost.

Guide Dogs and Bonds of Trust

Robert Deems is a fifty-one-year-old attorney who has been blind from birth. For many years he has used a guide dog. By law, such an animal is considered a part of a person and can go anywhere, and Robert and his dogs have developed a special degree of closeness and a strong mutual trust.

His office is on the tenth floor of an office building in downtown San Diego, and he lives about 6 miles away. His girlfriend of four years, Cathy Momile, drives him to work or he and his dog take the bus. The courthouse is a short walk from the office.

ROBERT DEEMS: I was born in McFarland, about 31 miles southeast of Parkersburg, which is right on the western bulge of West Virginia. I started going blind when I was just a baby. I had an unusual type of glaucoma. None of the doctors discovered what the problem was. It probably would have been treatable had it been discovered early enough, but in West Virginia doctors were not that great. Also, my family was very backward. They were mostly farmers. My father could write his name and that was the extent of his literacy. My mother had just gone to the fifth grade but was very bright.

She got us all interested in reading. Our entertainment on winter evenings was sitting around the dining room table with her reading us books. She read everything from the *Iliad*, the *Odyssey*, and western books. There are four brothers and five sisters living. Mother died when I was thirteen, and I went to a residential school for the blind. I would go in early September and stay through May. Then I went to the Ohio School for the Blind. Then for many years I lived in Ohio and had a riding academy. I rented out, boarded, bought, sold, and traded horses.

After I was divorced several years ago, I lost confidence in myself. (Before that I always had used a cane or nothing.) I decided that a guide dog would be the kind of crutch I needed. A blind person more than any other must have confidence in himself or herself. The old saying, "He who hesitates is lost," is particularly true for us. If you hesitate, everybody around you is going to lose confidence in your ability to do whatever you are going to do. People told me that guide dogs did prevent them from running into things, from going into holes, from stepping out in front of moving vehicles, and things of this nature. I decided to try one.

81

First you pick the school. There are about eight or ten of them in this country. They normally write back and ask about your medical condition and for references. Then they accept or reject you. You must be physically able to handle the dog. If you are a beggar they will generally reject you. If you are an alcoholic or drug addict they will reject you. If you are mean or unreliable they will reject you. Since I had gone to the Ohio School for the Blind, I knew about the Pilot Dogs in Columbus. At Pilot Dogs they are a little less formal than at other schools. They emphasize the dog being friendly, outgoing, and getting along well with other dogs as well as people. The training you go through isn't that complicated, but it is a whole new ball game. They are training you to let the dog lead you rather than you lead the dog. But you also have to pick up the slack if the dog makes a mistake. You have to remain alert. It's teamwork.

If you take hold of the handle, you can feel exactly what the dog is doing. A blind person is already used to having been led by a person, so it isn't entirely new. The dogs have a high level of understanding. One fellow told me that once he went into a post office to have his paycheck cashed. He missed his pocket and dropped his wallet without realizing it. When he got home, his dog had the wallet in his mouth. If there are places where dogs don't like to go, they can try to lead one past, but usually they give themselves away by looking at where they are supposed to go. A thing that is really funny is, it can take them about ten minutes to get home from work, and to go to work it can take about forty-five minutes if the dog doesn't like to go.

I got my first guide dog in March of 1965. At Pilot Dogs, the first two days you don't have a dog. They have you walk with the trainers so they can match your walking speed and your personality with the dog. The third day the director of training said, "Well, I am going to give you your dog today." He walked into the grooming room and he let Gretchen in. She came over to me, and I must have gotten this look on my face when I started petting her. I said, "Is she a Dobie?" I had wanted a German shepherd. He answered, "You try her for a few days, and if you don't like her I will give you a different dog."

I picked her up, put her in the sink, and gave her a bath. All these stories about how treacherous a Doberman is were on my mind. I took her down to the room with tile floors and started drying her. Later I learned she had a playful habit of clicking her teeth. But at that moment when she clicked her teeth together right in front of my face I thought she was snapping at me, so I just whammed her, and she went flying clear across the room, her feet scrambling on the tile. She let out a playful growl and came running back to me. So I realized she was just wanting to play and we started to wrestle. We made ribbons out of the towel.

82

Her previous blind owner had had cancer of the larynx and died. She had been brought back and retrained before I got her. She had gone out with him at about fourteen months and was returned about ten months later. She was very nervous about my going downstairs since that fellow had been very weak and had had difficulty going up and down stairs. She was afraid that I would just disappear too. I realized that I had won her over when the director and I were in the recreation room. We were bumping each other, trying to push each other off balance as men will do. Gretchen thought he was trying to hurt me and went for his leg. We had to spank her for that because she was being too protective, but it made us both feel good because we realized that I had won her over. It had taken about three weeks.

In December of 1965, we moved to San Diego. I had always wanted to be an attorney. In California I learned of a program called Aid to the Potentially Self-supporting Blind, and I began City College. I made pretty good grades and the rehabilitation counselor for the blind jumped on my bandwagon. After that, it was pretty much what you might call smooth sailing. I went two years to City College, then finished up at State in June of 1969. I started in at the University of San Diego Law School on a scholarship, graduated from there in 1972, and took the bar examination that year. Then I started my practice.

I had Gretchen until 1972, when I was studying for the bar exam. Someone gave her strychnine and she died. We had gone home from the bar review one night, and after I let her out in the back she acted funny, but I didn't think anything about it. The next morning she had dumped on the floor and was acting sick, but I didn't realize she had been poisoned. I had a hearing for a client, and when I returned I took Gretchen to the vet, but it was too late. I had them autopsy her and they could tell it was strychnine. It was like someone had murdered a good friend.

My classmates had felt like she was a member of the class. She had gone to college and law school with me. As soon as it was time for the class to end, she would jump out in the aisle and start looking the teacher right in the eye and go "Hoooooo." He would say, "Okay, you can go." She and I would be the first ones out the door. They all called her "the fifty-minute dog." For graduation, they gave her a degree, a Ph.T. (Putting him Through). When the fellow who was handing out the degrees called her name, she went up. They all loved her. It was exceptional, the way she would perform.

I could take her into a building and tell her to go to the telephone, and if there was a booth she would take me to it. I could tell her "escalator up" or "down" and she would take me to them. I could let her go anywhere with my son and I didn't have to worry at all. There was no way anyone could have bothered him. She would have torn them apart if they had tried.

There is a special level of trust that one develops in these animals. In 83

1969 I was going through a lot of stress, and I was having the final test of the semester at State College. I was going over to a friend's place to review his notes. As Gretchen and I turned into his apartment building, I was thinking about the test. Gretchen stopped. So I took my right foot and put it out. That is what you are supposed to do to see if there is anything there. I didn't find anything, so I told her "Forward." She turned left and right, letting me know we could turn around and go back but we couldn't go forward. I thought there must be an extension cord or a hose across the walk and she was afraid I would trip over it. I said, "You dummy, forward!" *Dummy* is a punishment word, so she went forward. If I had stayed close to her we would still have been all right, but I didn't. I took a step on my left foot; then I started to take a step on my right foot and there wasn't anything there. Suddenly I remembered: swimming pool!

I knew it was either jump in and go down feet first or go down on my side. I jumped sideways and landed feet first in the swimming pool, and of course I pulled Gretchen in with me. This poor woman came running out of her apartment, yelling, "Stay right there, stay right there." I was breaking up. I took hold of the side of the pool and swung out. Then I reached down, and old Gretchen popped both front feet in my hand and I lifted her out. I became famous throughout the state at that point. Whenever I would meet a blind person, they would say, "Oh yes, you are the guy who fell in the swimming pool with his guide dog!" Gretchen was so mad at me. She wouldn't have anything to do with me for the rest of the day.

Gretchen was poisoned in the middle of June of that year. I would sit there and reach down to pat her 'cause she was normally at my side. She wasn't there and I would break out crying. I couldn't help it. I had her cremated and her ashes scattered. I want the same for myself. I'm not one for memorials.

Having a guide dog was a plus and a minus. It brought back the confidence. In taking on a crutch, though, I developed a real need for the crutch. I have never been able to get around nearly as well since I have had a guide dog as I could prior to getting one. It has made me a more dependent person, and I don't like that part, but I do like having a dog with me at all times. They are an icebreaker, a conversation piece, and they really carry their own weight. For that reason I am not too unhappy that I got one. But blind people travel very well without a guide dog. In fact, only a small minority of them have guide dogs — I would say no more than 5 or 6 percent.

I arranged to get Ginny shortly after Gretchen died. At the Pilot Dogs School they wanted me to get another Doberman. I said, "No, I would compare it too much to Gretchen." The first time you go it takes a month; the second time it is two weeks. They expect that you are properly trained and that you will be able to handle the dog. Ginny is a Vizsla, raised by a lady in Iowa and donated to the school when she was six weeks old.

Ginny is great in court. She will lay under the council table all day long and just sleep while we are in trial. She is protective, too. If my girlfriend and I are out walking her at night and someone comes up behind us and starts following us down the street, she will look back over her shoulder and go "Wuff, wuff." One time these two great big German shepherds charged us, and she jumped between us and started to fight both of them. If they came at her and she was by herself, she would flop over on her back and wave her feet in the air and try to talk them into not attacking, but when she feels they are going to attack me or someone she cares about, she is ready to die to protect. It's a wonderful feeling to know that something alive would give that life for you.

I really don't think that Ginny understands my blindness. She will stand and look at me when she wants something. She expects me to see her. Of course I normally feel her being there, so I will ask her what she wants. When I hit what she wants, she will jump and take off either toward the door or the kitchen or whatever. But she will play a joke on me. I will call her, and she will get just out of reach and stand there with her tail wagging, like she is almost smiling at me. People have told me she does this.

I have been worrying about her recently. She is ten years old and is developing cataracts. She will go blind eventually. I will of course keep her as a pet. I would never allow her to be destroyed. She now has problems when we are going toward the sun. I have to help her along at times. I think the light hurts her eyes.

This is a funny story. One day we were on a corner, waiting to cross the street. A friend of mine was with me. This lady came up and whispered to him, "Is that one of those blind dogs?" "Oh yes, ma'am, and that gentleman is trying to help it across the street." I broke up. We had already crossed, and she was still standing there and thinking about what he had said. I thought that was good.

Dolphins and the Quest for Interspecies Communication

Dr. John Lilly is a controversial figure. Depending on one's point of view, he can be seen as a neuroresearcher, an expert on dolphin vocalizations, an ecologist, a prophet, or a mystic. After training as a medical doctor at the University of Pennsylvania and doing research in the neurophysiology of macaque monkeys at the National Institute of Mental Health in Bethesda, Maryland, he began to study the brain of the bottle-nosed dolphin, *Tursiops truncatus*, in 1955. He and his wife, Toni, believe that the 85 lessons to be learned from communication with dolphins will help humanity to abandon the vainglorious notion of being a "COU" (Center of the Universe).

After years of interaction and research with dolphins, the Lillys have become so convinced that they are fellow consciousnesses that they advocate protective laws comparable to the codes governing treatment of fellow human beings. They have, for example, suggested that dolphins involved in research settings be provided with telephone links to the open ocean in order to communicate with their fellows and families.

John and Toni live in an art-filled, elegantly rustic house in the Malibu hills. Their work with dolphins, however, is currently being done in Marine World in San Francisco and Careyes, Mexico, to which they commute frequently. After I met the Lillys, Toni suggested, "Why don't you swim with the dolphins? That's the only way to really understand what we are saying." So I donned wet suit, mask, and snorkel and entered the research pool — two large, circular tanks joined by a narrow channel. Immediately two dolphins swam to me, and I was impressed by their alertness and attempts to interact. They glided alongside, allowing me to stroke their warm, smooth skin. Quite rapidly they taught me a game: One would come up on either side of me, allowing me to grasp a dorsal fin in each hand, then they charioted me around the pool under water. I was left with an unforgettable sense of their intelligence and gentleness.

JOHN LILLY: I went into investigating the cetacea blindly, with *Tursiops*. They became my friends years later. But in the beginning I was looking on them as objects to be examined and brains to be cracked open and probed. While I was doing that, they kept doing unexpected and new things.

My turning point occurred in 1961. I listened to a tape on a friend's recorder at half-speed, and suddenly I heard the dolphin mimicking my laughter and mimicking electronically generated tones in the room. I got a very weird feeling that this must be something very bizarre. About a year later I said to my assistant Alice Miller, "Let's try it with Elvore." So I went in and shouted, "Elvore, *squirt water*." He came back with "Squeert wadde, squeert." I said, "No, *squirt water*." So we both worked on it, back and forth, back and forth. And finally he got it. It was so loud and piercing. It was incredible. I felt suddenly that I'd been projecting an awful lot of stuff onto dolphins from my being human. Suddenly all of that didn't mean anything. He himself was really there, the way another human is there for me. It was an incredible experience. Suddenly all dolphins became "people."

Dolphins have voluntary respiration. If the blowhole is in the air they can breathe, if it isn't in the air they can't breathe. That's very demanding. However, they spend 99.99 percent of their time underwater, so they are really water beings. This affects their consciousness. When you depend on the air above you for sustenance, for consciousness, for continuing the trip, you make sure that your friends are with you, either above you or below you or at the same level you are. In case anything happens to you, they can bring

you quickly to the surface. This means interdependence, which means communication. Without it, the dolphins wouldn't have survived the fifteen million years that they have been here on this planet. And they use the air underwater, force it back and forth between sacs, to make their sounds. Though the anatomy of the air sacs had been described, people didn't realize it was for vocalization.

In the sixties, we used nonsense syllables, and they could repeat ten in a row without any problem. If you put in a word like *correction* and started your sequence over, the dolphin would leave out the part that came before *correction* and give only the part that came afterward. So they began to understand our language, and there was a set of instructions on how to handle sound. The key was working with them in shallow water. For instance, Margaret Hone lived with them in 19 inches of water, day and night, for months.

Dolphins are very strange and alien, of very high intelligence, of very high sensitivity with a good deal of empathy, and with amazing attention to detail. They can follow hand gestures, body gestures, and contact in the water. There's a whole set of languages, paralinguistic things, that they understand and that they teach everybody who swims with them. And they teach unequivocally — they do minor punishment, they do reward. But your belief system determines what you see, and there is where the problem lies. After all, they have been around for fifteen million years with brains the size of ours, so their knowledge is old, alien, and in another element, a thousand times as dense as air. They use three-dimensional movement. They have to go to the ceiling every time they want a breath. They do have an awareness of our emotional states. They look for signals. They want to dissect what we do and interpret it in the form of reproducible signals. Anything you introduce in your relationship with them is interpreted. They do want to make contact. In studying them, we have to watch our humanness, we have to watch our abilities to project our own thoughts and emotions and doings onto other animals. But even when you attempt to allow for that, and attempt to subtract that, you still are left with a very large residual order of high sensitivity, intelligence, and intense attention.

Well, everybody who swims with them gets a feeling of rapport and of the attention that the dolphins pay to them, either in a positive or negative way. Some people who tend to project negative things if bopped by the dolphin think it is a hostile action; however, other people who are much more broad-minded think of that as a statement by the dolphin. When people do that kind of work, you begin to see communication. One man, who got bopped five different times in five different ways, came back and said, "You know, it is like a sentence. Each time it's a different sentence." I think the first communication of man and dolphin has to be a creative combination between us. We must add things, they add things, and so on. So we will develop a third language; it won't be dolphinese and it won't be human. I believe that the yearning to make contact is in their genetic code. They have been approaching man every time man has given them half an opportunity. Aristotle wrote about it two thousand years ago, and numerous other people have written about the contacts between humans and dolphins.

Our computer work with the sounds formed in their frequency range, which is ten times ours, is moving along. Our problem is getting them to match those sounds well enough so that we and the computer can recognize that they are matching. That's always the problem in communication. Can you match sounds, words, sentences, and so on, until finally you can exchange ideas? If you built houses on the edge of the sea with waterways of eighteen inches of water, the dolphins could come in and look at what the people are doing all the time, day and night. Then they would probably develop this third language. We visualize having human mothers in these communities giving birth to their babies underwater (which of course is the safest way to be born; the nicest way to be born). Being born with the dolphins would be great. The Russians have already done it. They have had pregnant mothers who are Olympic swimmers giving birth underwater with dolphins present. The human beings had no panic or fear. The presence of the dolphins dispelled all that. The dolphins interact with the babies, and they bring them to the surface like they would their own. The Russian movies of this have been shown on TV. You could develop these human-dolphin communities at the edge of seas all the way around the world; it would be a great international network—the Greek islands, in Hawaii, the Seychelles. I think it is possible.

The notion of animal rights came to me the day that Elvore said "squirt water." There was a "person" there. Many people can understand. The young ones do, those that have taken acid do, because their minds are open to alternatives. We have the term *eternity*. There is another term that I would like to introduce which I borrowed from science fiction, called *alternity*, the alternate ways of looking at things and the alternate states of consciousness. *Alternity* is a way to look at all of the possibilities, all of the probabilities. In the future I hope the youngsters will take it up and expand it.

TONI LILLY: I think communicating with another species will be our salvation. Because when you feel that you have the only mind, then the rest of the world is mindless. Then you don't have to use ethical or moral considerations and you can exploit and pollute. But if you feel that there is something there with a mind, then you and they are equal. Man is like that. I think that is our hope: to understand the interdependence of all things on the planet. Maybe we aren't supposed to last. Maybe that is in the order of things too. 89

But I would like to give it a try, and I want to put my effort and my time into something that makes some difference.

When John and I first met, he had just gotten back from Arica, in Chile. He had gone down to an esoteric school to learn a little more about himself because he had come to an impasse in his own consciousness. We met at a party, and it was an immediate recognition of each other. It wasn't an explosive thing. It was just an all-absolute knowing that we would be together. I was forty-three. We have been married eleven years.

We didn't talk about dolphins at all when we met. After about four years we decided to go back to dolphin research. It was a series of partial frustrations as I began to understand how a scientific field operates with dolphins. I began to see the problems with raising enough money to do the research, and I have gotten embroiled somehow doing an awful lot of that. With all that hard work, though, I think we've all accomplished something that is now at a point where we can go somewhere with it. The human-dolphin community is the thing that really interests me, because I think without a question that the real breakthroughs are going to come with children as they are forming their language and dolphins as they are forming theirs. They will be the real space children, because they will develop the thinking that these other creatures are sentient beings because they *are* sentient beings. The dolphins have a complicated acoustic vocabulary, and the human beings will be able to overlap and develop larynx functions that can make those sounds. John's daughter Cynthia heard dolphin sounds, whistles and clicks, at eighteen months. She can make sounds that no other child can. If you feel that there's just you as the superior one on the planet, there's only one mind and you're it, it is a very lonely place to be.

Cat Therapist

The bedrock of the practice of psychiatry is the belief that people have inner lives, that their symptoms are manifestations of stress from their present life or past histories, and that a therapist has access to understanding by training and by empathy. Problems are eased or cured by providing an atmosphere in which stress is lessened, so the patient can have the opportunity to learn new coping mechanisms. Sometimes medication is prescribed for relief of symptoms while the patient tries out new behaviors.

Carole Wilbourn, a therapist, bases her practice on psychiatric principles. Her clients, however, are not humans but cats. And she has shared her knowledge with 91 the public in her books *Cats Prefer It This Way, Cat Talk, The Inner Cat,* and *Cats on the Couch.*

Carole shares her charming, Early American antique–filled Greenwich Village apartment with Sunny Blue, a Siamese, and Honey Blue, a tortoiseshell cat.

CAROLE WILBOURN: I feel so complete when I sit down and write about cats. But what I really like to do the most is answer people's questions. I get letters from people who have read my books. They describe problems and I just sit down and dash out an answer. It's very encouraging. Sometimes I get phone calls. I would really like to have a syndicated column because it would be more immediate.

I wrote my books to try to teach what to do when a cat is uncomfortable and the person doesn't know what is wrong. Cats' behavior is the way it is for a reason, because of what they need or what they feel they need. Humans intellectualize their feelings, then spend twenty years in psychotherapy, but cats don't do that. They are more in touch — less sophisticated, but more in touch.

When I was four or five I got my first cat, Whitey. I have many bad memories of my childhood. There was never any intimacy in my family and my parents didn't really care. They weren't animal people. I had a sister who was ten years older, a brother, six years older, and a sister five and a half years younger. I was the middle. They got a dog when my brother was a little baby, but he was frightened of it so they had to give it up.

My memories are very difficult. My mother would sometimes tie the cats when she didn't want them to roam around when she was cleaning. She would call the SPCA to come and take the cats. I don't hate her, but I hated her obliviousness. She just didn't have the time or understanding. When I was nine I got my first dog, but each of my animals "disappeared" or just ran off. The animals made me feel good because I could talk to them. I had a canary when I was about nine or so, which I had for many, many years, and I taught it some tricks. Everybody had parakeets, so I used to teach the canary to sit on my shoulder.

I went away to school at seventeen, graduated from NYU, and went to live in Greenwich Village at twenty-one. Then I got a cat through an ad in the paper. He was supposed to be part Siamese, but he had tiger stripes and green eyes. I named him Oliver. I got the next cat, Sam, for my first husband as a birthday present before we married, and I got him after we divorced. Sam made me feel really good. I probably felt that I never got enough attention and so my heart went out to anything that needed love.

My mother told me some years ago that when she was pregnant with me she almost decided to have an abortion, and when she was on the table she changed her mind. So I deeply identify with unwanted cats and have this great need to nurture. I can also understand the pain of lack of nurturing. It gives me almost a firsthand experience of what it must be like for a poor pet animal because of the freedom people have to give it up or to keep it. Animals pick up rejection from people's bodies, so they are aware that their lives are precarious.

Cats can feel happiness, sadness, anxiety, whatever. The emotions affect his personality, his thinking, his doing, his being, and his spirit. If any of the emotions are traumatic, they can cause his behavior to become bizarre. So in figuring out why behavior is bizarre, it is a matter of deciding what was traumatic for him or her. You can make things better if you can get to the core and change the situation, the environment, or whatever stimulus is causing him to be sad, anxious, or mad. Sometimes it's impossible. If you have a cat and person who just do not get along and the person can't change, then the environment can't change and you have a static situation.

I see myself as a cat therapist. I try to help people understand their cat's day-to-day behavior. When things change it is because there is a reason. Unfortunately, I don't get many cases until the problem is serious. But it's ideal if I have people who have an older cat and they want to adopt a kitten, and they ask me what kind of kitten they should get and how to go about it. Then I can help them right from the beginning. But many of the cases are after they get the kitten for the cat, when they are having problems. For example, if you have a two-year-old male who is aggressive and very social, I wouldn't get him a shy little kitten; I would get him a spunky, probably female kitten because their energy is softer. If I had a ten-year-old cat, I wouldn't get a bouncing kitten. I would try to get a wise little mellow cat. If I had a little kitten, then just getting any kind of a cat would work out.

I started seeing cat patients at the end of '76. First I called myself a behaviorist, but people thought of Skinner, and the connotation wasn't right. The media kept calling me a cat shrink. I didn't have any formal training as a cat therapist, since there isn't a cat therapist school. But I majored in psychology, I have taught, and I have worked extensively with cats.

I make house calls, and I feel especially good if I can help a cat. For example, I was called to a family where their Siamese wasn't getting along with a new kitten. I realized that the mother and two children, eleven and sixteen, were going through separation anxiety because of the mother's recent divorce. The Siamese had been in a recent accident, an attack by a raccoon, and his coordination was off. He felt he belonged, and he identified with the older girl. The new black kitten identified with the younger girl. The girl was having a rough time because her father wasn't there, and it was in the midst of her separation anxiety that she had brought the black kitten home. Of course, all the attention went to it.

The Siamese just couldn't deal with the newcomer and all the anxiety going on in the household as well as with his coordination problems. I tried to tell them why the Siamese felt the way he did, and that there was a possibility he would never accept the black kitten. They had to keep the two cats separate. This was difficult because, since the father left, the two girls slept together with the mother each night. I told the mother to start the Siamese on Valium as an ancillary support to relieve his anxieties. The primary support was giving him a lot of love and attention to build his self-esteem. If he didn't feel better or if the cats could not be kept apart, then it would be necessary to find another placement for the black kitten.

I used my understanding to select a companion for my Siamese, Sunny Blue. I had gotten Sunny as a companion for my cat Bagins, and Bagins passed on a year ago this November. From November until July I really didn't have a companion for Sunny. He wasn't that secure with himself, and he needed the time to feel that he was Number 1 'cause he had been Bagins's companion. Originally he was found on the street, and I don't know how many times he was abandoned before that. I was trying to make him feel strong within himself, to let him know that each time I went out I would come back again, and of course I took him away with us whenever possible. The hope was that in the future a companion could come in and Sunny wouldn't feel displaced.

Around May I thought that he was feeling pretty strong with himself, so I called up the girl who had originally found Sunny. I asked if she could find a kitten. I thought it should be a tortoiseshell, because Bagins had been black and white and Sunny was used to the dark colors. Tortoiseshells are usually small cats and Sunny is a small cat. Also, tortoiseshells sometimes have Siamese mixed in them, and that would be a good feeling for Sunny. So I needed a cat that was sort of fragile and light in energy, and it seemed that the tortoiseshell type would be it.

It took her a few months, but she finally found one through a veterinarian in Brooklyn, who told her about a woman who had given a tortoiseshell kitten away. She had found her by a garbage can along with her little sister when they were about three months. She had given her to a woman who had to move and returned her. That is how my friend got Honey Blue.

When she brought Honey Blue to our apartment, we kept her in the carrier for a while and then my friend opened it. Sunny was content just to lie there. Finally I distracted him, and when he wasn't looking, my friend let the kitten out. Honey went up on the seat for a while and sat there. Then Sunny went over and sort of looked at her. She ran into the bedroom and he ran after her. There were a few skirmishes between them — nothing really bad. I told him that things were all right and that she was going to be his honey, which is why she was named Honey Blue. Within a couple of days he was washing her because he is such a poppa and he loves to lick.

He takes care of her now, but he has to know that he is first because of the insecurity he has gone through in his past. He just sort of relives everything. Whenever I talk to her, I say she is Sunny's Honey so he feels included. I try to pet them at the same time so he won't feel that I am going to throw him out 'cause I have her.

It comes down to this. In trying to heal a cat, it's like getting in touch with the essence of its life. If you can get at the core of a problem, there's a chance it can be solved. Life just goes around in a circle. We give and take from each other. I can try to remove the blocks and keep the harmony going.

Snakeman

Parked in the driveway of his modest ranch-style house in a crowded suburb of San Diego is Bob Applegate's green Ford van. A cartoon serpent labeled SNEAKY SNAKE is emblazoned down one side, and the words "SLOW — LADIES LEG INSPECTION AHEAD" are artfully lettered on the rear. Bob, a thirty-seven-year-old fire chief, has kept snakes for most of his life.

One of the three bedrooms is called the Snake Room and has custom-built, wood and glass breeding cubicles, each with an upper level and a lower drawer for a breeding chamber.

In the family room is a large wood and glass cage. The large pythons and boas live there. Outside is the Snake Pit, a concrete, walled area 4 feet deep in which ponds and rockwork provide a natural but escapeproof setting for snakes and lizards. The garage has been partitioned into cages for housing additional snakes. Bob has about one hundred snakes in all, including many species of rare king snakes, and pythons, boas, rat snakes, gopher snakes, and bull snakes. He is a founding member of the San Diego Herpetological Society and writes for its monthly newsletter. He attends reptile breeding symposiums and gives lectures at schools and to various groups.

BOB APPLEGATE: Well, from what my parents tell me, I was the type of kid who would like something that everyone else would be afraid of. I thought that all snakes were dangerous. At around age six I found this snake about a foot long. Somehow I managed to put a tuna can over it and slip a board under. During the process I got bit and thought, "Oh God, I've had it." Anyway, I didn't die on the way home, and I got up the nerve to hold it. First my parents wouldn't let me bring it in the house because my mother was terrified, but finally they let me put it outside in a ham can with a board lying over the top. That first night it got away, so my dad made me a Plexiglas cage. The next snake I brought home was a striped king snake, and Dad 95 managed to get permission from my mother for me to keep it in the house now that I had this cage that nothing could get out of. Then I had fun running around the neighborhood, scaring everybody with my snake.

My dad was tolerant, but he didn't actually like snakes. He realized it was a beneficial way to channel some of my energy. We moved when I was ten, and I brought maybe half a dozen snakes with me, which more or less took over my bedroom. My mother by this time had become tolerant, and I think she actually kind of enjoyed them.

Somewhere between the ages of ten and fourteen, I traded off a bunch of baby snakes for a boa constrictor from a man doing research at the zoo. It wouldn't fit into any of the cages, so my mother and I agreed that if I kept my bedroom door closed, it could run loose. When my mother went in to clean the room, she would pick it up and set it on the lawn outside the window where she could keep an eye on it, then bring it back in. Still, things happened that she didn't like. I used to catch alligator lizards and send them to a guy in Florida and got 50 cents apiece for them. I made pretty good money for a kid, but he only wanted them in groups of fifty. I remember I had this boxful. The water must have spilled, the box collapsed, and all the lizards got loose in the house. For weeks you'd have to pick up your clothes and shake them, and every once in a while a lizard would fly out. By that point I had cages mounted on the wall instead of shelves. My dad used to help me build them. He would bring home packing crates and we would buy glass and stick it on.

My father wouldn't let me keep poisonous snakes, but you go through a stage when it's thrilling to have them. I caught my first rattlesnake when I was twelve or thirteen. To catch them, I would get a stick and pin their head down and grab it and show the fangs and let the venom run down my fingers. It was all thrilling and a macho-type thing. But I started realizing around then that snakes are living things too, and if you wanted to keep one you should try to be gentle. Because if you pin one down real rough and handle it real rough it will go into shock, and it won't eat and will starve to death.

Well, most kids grow out of keeping pet snakes. I'm sure my parents hoped it was a passing fancy, but even then they must have realized that I was getting into it a lot deeper than the normal kid. But the hobby kept me hiking around in the hills and generally out of trouble, and a lot of my friends were running around in little gang wars and going on to bigger things. I think nowadays I would have been categorized as a hyperactive kid. I remember in third grade I was so much trouble that my teacher said, "I don't want you in my class anymore. Go down to the next room and ask the teacher if you can be in her class." To this day I remember it 'cause I was actually afraid to do it, but I didn't want everyone else to know I was afraid.

There was a short break in snakekeeping when I got married at eighteen. I was maintaining a full load at college and working two full-time jobs and a part-time job. Some friends of mine were going down to Mexico to go collect snakes, and I wanted to go so badly I could cry, but I had to keep working to support my family. They brought me back a baby boa constrictor, and that started the snakes again.

Well, with a daughter and a wife living in a small apartment, it took a while to get back into it. I gradually ended up getting a little bit of extra money, and I didn't have to work two or three jobs. We did have to live through almost getting evicted when the landlord found rattlesnakes in the bedroom. I was married for twelve years but have been divorced for several years now. My wife had been tolerant. She would have said, "If he isn't chasing other women, snakes are worth it."

In 1970 I moved to my present house and put a snake pit in the backyard in 1973. To conform to the city ordinance, I petitioned to have my snakes considered ordinary household pets, and the city council agreed. Actually, I'd be kind of afraid to leave the area because I've kind of established my legal precedence.

As a kid, I wanted to figure out a way to be in reptiles for my living. The only financially rewarding position was to be a college professor, but I wasn't able to afford the years of schooling. Zookeepers weren't paid that much, and you almost needed a doctorate to become a curator. Now, my dad was a fireman, and I was pretty familiar with the schedule. Most people work a forty-hour week from eight to five, but a fireman goes in for twenty-four, with eight hours work, eight hours standby, and eight hours sleeping time. Then you get to stay home from eight in the morning until eight the next day, which gives a lot more usable time. They rate the job probably as the most dangerous large profession. But I love it; I wouldn't trade it for an eight-to-five job.

Every two weeks I buy wholesale between one and two hundred mice to supplement the ones I raise, a hundred baby chickens, forty to sixty rats, and rabbits as available to feed my snakes. In the last couple of years I have seen the handwriting on the wall. In the past, if a snake died, it was always less expensive to write a letter to Thailand, the Soloman Islands, or India and ask for a shipment of them. The zoos were like this. Everybody was like this. Now you can't do that anymore, so if people are going to continue to have reptiles, particularly the exotic ones, someone will have to breed them in captivity.

My goal is the propagation of certain species. I feel that if the wild traits of an animal are compatible with captivity, an animal can be kept in a house. Snakes have very few basic needs, and these are easily met in captivity — you feed them, give them water, keep them at a certain temperature, and they can go for long periods without handling. If the situation is correct and you have the right sexes, they will breed.

I make my cages so they are a real nice place for the snakes to live. I actually feel that most of my animals live longer than in the wild, since I protect them from predators and eliminate parasites and treat illnesses. I have very good breeding success. A few years ago my Argentinian boas, which I had gotten before they became a protected species, had babies, first in June, then in September. To my knowledge, these were the only two births of that species in the United States at that time. Currently, I am breeding many varieties of rare king snakes, like Central American milk snakes, gray-banded king snakes, and ones that look like coral snakes. I have a high hatching ratio. I had sixty-eight California king babies. I hatched fourteen or fifteen mountain kings, innumerable gopher snakes and glossy snakes. Hatching kind of keeps you on edge; it's good practice to keep your techniques up. I've had eggs from gila monsters, ball pythons, Burmese pythons, miscellaneous African snakes, Western milk snakes, hog-nosed snakes, black rat snakes, pine snakes — almost everything I've had has laid eggs that I've managed to hatch.

I also observe matings and check the cloacal openings of the females for sperm samples, which I check under the microscope to see if the sperm is viable. I prefer to keep track of who the father is, so if I do not have a viable sperm sample from one male I introduce a second. This can be done since the female is receptive to a number of males. I like to keep a photograph on record of each of my animals. So if I trade or sell anybody a snake, the photograph can be like an AKC registration. I can show a picture of the parents, grandparents, and great-grandparents. Then, if somebody is interested in a head marking or a certain scale series or a line of blotches, I can show them pictures where they can trace back through the generations.

I have animals that are individual pets, even though I try to be practical and have animals that I know will be good for my breeding programs. The one who I will never forget is probably my big python, Reno, that died trying to have eggs this year. She was friendly; you would open the cage and she would crawl over to you. She was hatched in captivity. She was my largest at the time. I had her for several years, traded her from a guy over in Reno, Nevada, when she was about 50 pounds. That's why I called her Reno. She made the newspapers on several occasions and was always a favorite in school lectures. I would let a few teachers and a few kids line up and each hold a section, and I'd take a picture. She was an impressive animal. I felt sad when she died. It was a feeling like if you lost someone — all the potential that could come out of the relationship was gone. It may sound pretty cold, but I gave her to a friend, who skinned her. I think I would someday like the skin at least to show how big some of the snakes can get. But I did   99 go over to his house, and he had it draped across the back of his couch, and I said, "Gee, that's Reno, that doesn't look right."

Snakes' intelligence is rated pretty low. But I know they vary as individuals. I will have a batch of twenty pythons hatch out, and maybe there will be two that instead of just coiling up or making a little S in their neck like they are going to strike will just cruise over to you to see what is going on. Now if you have a good friend, that is the one you give him for a pet. I had a guy come to my place with a 12-foot reticulated python that he could set down in any part of the yard, and the snake would crawl over to him and crawl up on him. Apparently there was a relationship between the snake and the person. I thought it was pretty amazing, though I am inclined to feel that learning in snakes is more a conditioned response than intelligent learning.

For the record, except for a practical joke now and then, I don't go around trying to frighten people with my snakes. I try to educate people so they can appreciate the reptile and perhaps understand why the love of them is such an important part of my life.

The Song of the Guinea Pig

Peggy Fry, sixty-three, has been observing and raising guinea pigs, or cavies, for nine years, spending several hours each day with them in a tiny house behind her own. She has a hundred and fifty guinea pigs, and within her subculture of guinea pig breeders, she is a celebrity because in guinea pig shows, she has won more than a thousand ribbons, almost all for first place. She has enough Best of Show certificates to paper a room.

Guinea pig shows have three classes, based on age, and each class is divided by color. The Juniors are up to four months, the Intermediates up to six months, the Seniors six months or older. In the judging, animals can win progressively Best of Class, Best of Breed (there are six: Peruvian, Abyssinian, American, Crested, Teddies, Silkies), and Best of Show.

PEGGY FRY: I always had a sense of competing. My father was a competer. He was almost a millionaire. Up in Nova Scotia, where he was raised, he decided there was coal in them thar hills. So he went to the government and took out ninety-nine-year leases on a lot of land. He was a barber by trade, and he drilled and hit coal. He made all kinds of money hand over fist. He was a competitor. He had a boxing arena where he trained boxers himself, and he had a racetrack where he trained his own horses. He had beautiful

pit bulls, and he swam them in competition. The dogs didn't fight, they just swam for speed. And he had fighting cocks till the day he died in 1959.

I never had much time for animals while raising the children, but I used to compete in canning, embroidery, and artwork. But my children always had horses. My husband is a carpenter, and I have two daughters, three grandchildren, and one great-grandson.

I got my first guinea pig nine years ago from my granddaughter. I became curious about them and started digging in and finding out what they were. That's when I heard that a local Cavy and Rabbit Breeders' Association was being formed, and I got in at the beginning. I had success at shows right off. I'm a firm believer in health, cleanliness, and schedule, and just started coming up with good results. I took Best of Show the first year I was in it, and I learned a lot as I went along.

I got interested in the long-haired Peruvians because they're the biggest challenge. If I was a kid, I probably would have gone into the Americans or Abyssinians, but being old and stubborn, I wanted the challenge of the hardest coat and the hardest colors. Your reds, your blacks, your creams, your whites — they are all challenges. Broken colors are easier because you could breed anything with anything and it wouldn't matter.

My guinea pigs are huge — very, very bricky with big, bold eyes. I don't like ratty faces. When they judge Peruvians, they look for an even balance of very thick coat the same length all the way around, like a complete circle. The coats must be soft and fine like silk. Color is important, and their ears should be like rose petals. They should have a good crown — it's the area right over the shoulders. It's muscle. And I don't want any flab. I keep them in condition.

I know that many of them enjoy their coats. One was Show Off. She was black and white. She was shown till she was maybe two and a half and still was in very fine coat. But she had found a spot on her feed dish and rubbed her chin bald. Any bald spot disqualifies them from judging. She had won so many shows that I had begun to pull her out. I think it's very discouraging to a beginner to feel they never have a chance because another animal always wins. Frankly, I was also tired of her long hair. Her coat was over 60 inches across combed out. So I cut her coat off thinking I was doing her a favor, but I really wasn't.

The day I cut her coat off I put her in a pen with other retired sows. She absolutely went crazy, she was so terrified of them. So I put her back in a cage by herself, but she just sat in the corner and screamed, the minute anybody came near. It was breaking my heart to see her like that. She didn't understand, and there was no way of communicating. I saw her start eating her lettuce, but when I went back to give everyone water and turn out the light, she was dead. Nobody could talk to me for days. I blamed myself. She was such a prima donna. Oh, she held her head up in the air, didn't want anybody to touch her face; she didn't like anybody to feed or comb her except me. When she was being judged, she would sit up there very quietly and lift her head up so proud and look around at everybody.

The long-haired cavies grow hair at the rate of ¾ inch per month as long as they live, so when they are three months old, you have to start wrapping their hair if you intend to show them. I usually wrap the coats high and tight so they don't get the hair caught in their feet. I keep them very clean, doing them religiously every other day. I never brush; I always use a steel dog comb with the large end and no pulling. I comb them one section at a time, and roll it back up. The overall combing out is much harder because you have to separate the hair again. I have had some where the hair is so long, I have to wrap two paper towels lengthwise around each section. For bathing I use Johnson's No More Tears shampoo, and I dilute it in half with water.

If nothing happens to it, a cavy can live for seven years. But I've heard of a man in Missouri who has one twenty-one years old. He keeps it right in the house in a big cage in the living room. It's supposed to be enormous.

It was February 1980 when tragedy struck us. It had been raining terribly all winter. All we had was about an hour's warning that a flash flood was coming. We lost everything in the house. I had no guinea pig chow. The roads were flooded over and we couldn't get out for two weeks, so I crossed the floodwaters on foot with two trash bags in my hand. People on the other side took me to the store to fill the bags with lettuce trimmings. One time when I was returning I tried to cross with one bag over my back, but the river current caught me and the lettuce bag, and we both went down. So a man carried the bags across the river for me. My husband told me I was crazy. But my guinea pigs didn't miss one day of lettuce.

Since I spend so much time with them, I have had a chance to learn a lot about guinea pigs. The mothers usually clean the babies right after giving birth, and the babies' bowels are stimulated by the tongue licking. When they drop that first little turd the mother goes right on to the next baby because she knows that one is okay now. That's why if the mother dies and you are lucky enough to have some babies, you take a wet towel to simulate the dampness of the mother's tongue and start moistening the baby and massaging it to get that first little movement out. Then you have to feed them milk with an eyedropper until they learn to drink out of a bottle themselves.

Some of the mothers eat the afterbirth. Then they lick the babies clean and put them under them to dry. Babies are born with a full coat of hair, open eyes, and are up and running. Some mothers look like they are ignoring the babies because they are exhausted. In general, guinea pigs have very little stamina. Childbirth is a very traumatic experience to them. The fact that the babies are so big at birth causes the problem. Sometimes the mothers have to get on their hindquarters, hunch up, and reach their head down, and pull, pull, pull on the sac. If the sac breaks before the baby is born, the youngster can breathe in the fluid and drown.

The males and females are very affectionate with each other. When a sow comes into heat, the boar will swing his hips and make a very long purring sound. They jump with their two hind feet straight up in the air, the way a mule kicks. He shows her how beautiful and big he is and how lucky she is. He will just perform, and always with the constant purr of the love talk. They lick ears, and some are quite ticklish right on their sides. Incidentally, if you are trying to bring one up for a show and it's ticklish, you might as well give up because they can never stand still when the judge touches them.

With the boars, you've got to constantly check their penis for hair or sawdust getting up in it, because a boar alone in a cage will rub his tummy and stimulate himself. You can find hair that they rub on the hay cubes, and it's actually a form of masturbation. Every time I clean them, I pick them up and check their penis because they do pick up an awful lot of stuff. People who don't check a lot must lose them because infections can start easily.

When they mate, it's very fast and repeated. The male can repeat and repeat for at least four hours. Constant breeding, and I mean constant. He'll get off her and he'll love-talk her and he'll get right back on her in a period of sixty-five to seventy seconds. The males have a gland that produces a copulation plug, but I have only rarely seen one, and it's not a sure sign that you're going to have babies.

I've also seen close friendships. There were two silkies, a brother and a sister, who were the first two silkies I ever had born in my caviary. I kept them mostly together for close to five years and hoped they would breed. The boar was silver and the sow was silver and white. Then the female became blinded in one eye by running into a straw, and the problem traveled to the other eye. She became totally blind. When she couldn't find something, he would nudge her to it. It was beautiful.

And then she got awfully sick, and I was giving her sulfur. I didn't know what was wrong, except I knew she was old and she did have that eye condition. One morning when I came in to feed they looked fine because he was lying around her. I mean, he was always wrapping himself around her. But when I went back to water they were still in the same position, so I reached in to see what the problem was, and she was dead. He was trying to keep her warm, and when I took her out he was very, very agitated. He fought my hand. It was very unusual, but he didn't want me to move her. Ah, I guess it was two or three days later, he died. Just mourned and mourned and mourned. He just wouldn't eat or drink. Nothing. No lettuce, no carrots, nothing. He just sat there with his nose up against the bars. In fact, that's how he died. I had taken her, and he was waiting for her to come back.

Another breeder told me she thinks I have such good large animals because they are like children and do well with a schedule. That's why you don't hear all the noise and whistling like you hear in other caviaries. They know when everything is coming and they don't have to worry about it. They are creatures of habit. I feed them Purina guinea pig chow because that is the best there is. Then I always use rolled barley for conditioning and greens, fruits and vegetables every day. Monkeys, humans, and guinea pigs can't store vitamin C, so they have to be given it daily. In the morning they expect me to come in first thing and push their pellets down because the trays don't always self-fill. This is usually about nine-thirty or ten o'clock in the morning, as soon as I get finished with my dishes in the house. Then I go through and I pick up my lettuce pans and I wash them and set them in the sun for drying. When I come in from that, everybody who has an empty bottle is shaking it because now it's water time. After they eat, they play and take naps. At four o'clock they know that I bring them fresh ice water, and they start rustling around, stretching, yawning, and going to their water bottles. When I finish dinner at night, I go out to them and push the food down again and put the pans down for the lettuce. Then everybody stands in a circle around the pans with their heads up, waiting for me to drop the lettuce in.

I've heard them snore and they do dream. I have this cream boar that can go to sleep and snore up a storm, like a little pony. At first I thought he had pneumonia because of that rattle, so I grabbed him up, and of course he became unglued because he was sound asleep. He had no runny nose and I had my ear to his chest, but I couldn't hear anything. I gave him a dose of Sulfur-6 anyway, but he has done it so much since then that I realized he just snores. And they dream — they will twitch just like a dog. Sometimes they'll wake themselves up and they'll sit there startled, like, where did it go? They'll look so stunned.

They communicate. A mother, when her milk builds up, will start this very low croon — a broken-up chatter, but it's soft like a croon — that pulls her babies right to her to drink the milk and relieve the pressure. When they want babies to eat, they will nip them in the butt and give a little chirp, which sends the little one up to the dish. The mother will also teach the babies to sit very still until she makes a purring noise for them to come. When adults sleep, if somebody goes by and trips over them or touches them they will almost growl. Whistling is a signal and a sign of excitement.

The most unusual sound you'll ever hear is a weird noise that sounds like a chirping bird. I gave a male to a little girl, and it was chirping so much that her mother asked me about it; they tore their house apart looking for a bird before they realized it was him. He will chirp even when you look him square in the face. One night the family was playing poker and they heard this noise, and sure enough, Darth Vader was chirping away, and they got twenty minutes of it on tape. It went up and down, and I don't know what he was trying to communicate to anybody, but it was beautiful. It seems to run on forever, and if you are not interested in it, it is a very boring tape. But if you are listening, and if you know guinea pigs, you hear all these different expressions, and it's really something.

I am just spellbound by the song. I try to fathom in my mind what they are saying while the other ones are silent, motionless, and listening. I think it is communication, because I think they have something to say. And the chirps are very emphatic: way up high, way down low, little short ones, and long shrill ones. You can almost see a conductor's hand waving. I hear it very rarely, maybe once a year, altogether probably fifteen times. It really takes you out of whatever you are in. If only I could fathom it. What they are saying is terribly important to cavies, otherwise they wouldn't be paying the rapt attention they do.

New York City Honey and How to Meet One's Own Bee in a Manhattan Park

Knox Martin, sixty, a New Yorker, can recognize his own bees because he keeps a species of the European honeybee, which looks somewhat different from the American type. He is an artist, and bees and other insects are one source of inspiration for his work. His beehive, 3 feet square and 5 feet high, is on the roof of the building where he has his studio.

About five blocks away is the six-room apartment he shares with his wife of fifteen years, Rosemarie; Zega, an African gray parrot; the dwarf chinchilla rabbit Jasmine; Alexander, a 8-foot boa; and Cap, a Yorkshire terrier.

Knox currently teaches at the Art Students League in New York; he has been an assistant visiting professor in art at the Yale Graduate School and has also taught at the NYU Graduate School.

KNOX MARTIN: A sense of wonder has never left me. It just grows. The wonder that breeds a sense of quiet ecstasy. I was born in Colombia, South America, and grew up in Salem, Virginia. My parents had an appreciation of beauty and nature.

As a child, I had an attic filled with captive animals. I had wolf spiders, tarantulas, garden spiders, a little collection of jewels that were jumping

spiders, and many other tiny things. I also had a collection of small sharks, crabs, flounders, and snails that I would get in tidal pools in Sag Harbor when I was about eight. And there were always dogs, always cats — from pointers to setters, to cocker spaniels, chow dogs, Siamese cats, Abyssinian cats, alley cats, and a Great Dane that lived for fourteen years.

The thing that is amazing to me now is that as a kid I would see grass and it would be staggering. I would look at grass — grass growing, a tuft of grass — and the green would be so vibrant and the shapes of the grass would be so incredible. And I somehow realized that this miracle manifested itself out of the earth. I was the oldest of three brothers. One was killed in the war. Another, a remarkable person, is living in California. My father came from wealth. He was a painter and a poet, but he gave up an artistic career for one in aviation. He was one of the pioneer flyers and was quite famous.

Apparently when he was an art student at Maryland University he heard a strange noise outside. It was one of the early airplanes. And he decided that's what he wanted to do. His uncle had been putting him through school, and he had promised to finish his art studies in Paris. So he had to go to Paris, but when he came back his uncle bought him his first plane. He moved to South America from his home in Salem to attempt to open up aviation in new countries. He christened his airplane the *Simón Bolívar*. In those days he was considered almost a god in South America. He was the first man to fly to the Andes, the first to fly to Bogotá. I was the first one-year-old ever to fly, according to my mother, when he took me up over Barranquilla in Colombia.

My father was killed in an automobile accident when I was about six. My mother married a commercial fisherman, a completely different person but a great guy. The sense of wonder that I had as a kid worked into my art. I have a painting that I did when I was fourteen that astonishes me.

I flew to repeat my father's flight from Barranquilla to Puerto Colombia when I was an art professor at Yale. I was invited to come for the celebration of the fiftieth anniversary of my father's flight, and I said, "Yeah, not only will I come, but I'll make the same flight." And I did.

Bees were an early interest of mine. I was about fourteen when I read Fabre, a French entomologist who junked all of entomology because he didn't like the dry textbook stuff, so he rewrote it with a warm sense of connectedness. For example, he called a wasp "a pirate with a poison dagger." I was lucky that his was one of my early books on insects. Before that, when I was twelve, I read a boys' book that talked about how to catch wild bees and find their honey. So I just watched for wild bees in my walks through the woods, and occasionally I found them, but I could never find their hive. Then I began to dream of having a beehive. So I read all the stuff on them. There is an extraordinary development of the hive from early times. Oh God, they go back to prehistoric man. There are cave paintings of beehives and guys getting honey—the most beautiful painting of a guy on sort of a ladder reaching into a hive in a tree, and bees flying around. The early hives were of woven reed. Every time you took the honey out of the hive it would be destroyed. Then Lany Stroth invented the modern hive. It was a great development, built by a little invention here and a little invention there. There is a seven-framed and a ten-framed version. Each frame has 10 pounds of honey. There is one frame that is the "super," the brood frame.

I planned my hive for a long time. About six years ago I set up the hive on the roof of my studio. It's eight stories up. I sent away to Sears, Roebuck through their farmers' catalogue. You get it all knocked down, and by putting it together you understand each part. It's an amazing machine. In each frame you put a sheet of wax that's embedded in wire and has a print of cells on it. One frame is the brood part. My hive has three sections and a top. Sears sends you the bees. I picked out "midnight bees." They said they were very gentle bees, and they were sent from Florida. You receive a mesh cage with roughly forty thousand bees, and attached to it is the queen in a little box that is shaped like a figure eight. Each end of her box is plugged with sugar candy.

After you set up the hive you take the roof off, open the mesh cage, and shake the bees in. Then you take the queen in her holder and put her in. She needs the holder because unless she gets the odor of the hive, they will sting her to death. They cluster all around her and begin to eat their way through the sugar plugs. It is timed to take three days, and by that time she has the odor of the hive. They clean and feed her and then they create the drones, the males. Then she goes out on her fantastic mating flight, which covers several miles, and she becomes impregnated. She comes back to the hive, and the males are cast out. They die because they are not able to feed themselves.

The average worker bee's life span is about four weeks. They fly themselves to a frazzle. The queen's life span is up to eight years. I still have my first queen. She is impregnated for life in that one flight, and she lays thousands of eggs. I never saw her nuptial flight, but I watched for it as much as I could.

Each year they make more queens by taking a normal cell with normal larvae and feeding them royal jelly. They enlarge the cells and create queens. They also create many drones. Then suddenly the hive senses it is getting too large, and they swarm — fly off with the one new queen that is allowed to survive. The original queen and the workers stay. The swarm usually makes it to lampposts on Amsterdam Avenue, and they become a TV news item.

These midnight bees are tamer than most bees. I can handle them with my bare hands. Most beekeepers have a bellows with a spout on it, and they fill that with the worst foul, oily rags they can find. They light them and they

smolder, and the smoke is puffed into the hive. This makes the bees feel the hive is endangered, and they quickly eat honey to sustain them in their escape. But when they get filled with honey they become more passive and are so engorged they can't bend their bodies to sting you. I never have to do this. The only time I've been stung is if I inadvertently crush a bee.

I watched a beekeeper recently who had about fourteen hives. But he had no knowledge of what the smoke does, because immediately after puffing it in there he went in and opened the hives. He didn't wait for the bees to eat the honey. I said, "What are you doing?" And he said, "Well, that makes them groggy." I said, "It isn't true." He said, "Don't tell me. I'm a beekeeper."

I can identify my own bees. When I go walking in the Cloisters, I see my bees there. I see other bees too, but I can tell mine because their tails are darker. It's only about one mile from their hive. There are wildflowers in the herb gardens, crabapple trees, and catalpa.

Just out of curiosity, I tasted a little bit of my bees' honey the first year. It was delicious. But I prefer just to leave them alone. Eventually I'll take the hive to my daughter's property. The winter would be a good time to transfer them because they seal everything up.

Who's to say that the American Indians were wrong when they conceived of a hive of bees as one spirit. Sometimes when I'm with them I have a sense of some deeper order, something beyond the mind. Ultimately, I feel it's all connected. Anytime a person feels separate from so-called nature, from fish, or birds, or insects, or trees, one is fragmented.

Controversial Champion of Animal Rights

Since its inception in 1967, the Fund for Animals has had its office in a venerable building in Manhattan next to the famous Russian Tea Room and down the street from Carnegie Hall. Cleveland Amory, the founder, president, and chief executive of the fund, is the author of *The Proper Bostonians, Who Killed Society?*, and *The Trouble with Nowadays* as well as a vocal spokesman for kindness to animals.

The office of the Fund for Animals is huge, with ceilings almost two stories high. It has Cleveland's library, all the research files, and areas for keeping financial accounts. There is a staff of seven people, including his personal assistant, administrator, and secretary of the Board of the fund, Marian Probst. Photographs, artwork, and mementos cover any remaining wall or floor space.

There are two office cats, an old black and white male named Benedict and a black six-year-old female named Little Girl. A block and a half away is Cleveland's 111 apartment, where he lives with a big white cat known as Polar Bear, a former stray whom he rescued. He has a married stepdaughter, Gaea Leinhardt, with whom he is very close.

CLEVELAND AMORY: I was born and grew up in the Boston resort of Nahant. My mother and father liked animals, but it was my maternal grandmother who got me the first dog I ever had. She said I was plenty old enough to have a dog, and she was a very strong-minded woman.

Brooky arrived when I was about seven. I remember his coming by truck, in a crate. They undid the crate, and he just bounded out right at me, and I held on to him. He was an Old English sheepdog, and he had come all the way from Maryland. He was a champion. I had picked him out of a book of dogs, and my grandmother got him for me.

My grandmother certainly taught me, from the moment Brooky came, that I was responsible for him, and that when he was sick I was still responsible, and when he was unhappy I was responsible. She taught me that looking after an animal wasn't just pretty; it would be sick sometimes, and it would eventually die. She told me all that when I was seven years old. I loved the responsibility, because I loved the animal.

My father lived to see the Fund for Animals become pretty large, but my mother died a couple of years before. I think my father never quite thought that what I was doing was important. He was in the textile business. But if other people would criticize me, he would be mad and very protective. I think my mother took about the same attitude he did, feeling it would be better for me to go on with my writing, which was going well, and that the fund was not that important. It was a hard thing being a humorist and having such a serious cause as animals. But I think my careers add up to an appreciation of the irony of mankind.

With every animal person there is a turning point, something that you someday see or that happens, and the spark of compassion that all people have gets lit. In my case, it was a bullfight. I was writing for the *Arizona Daily Star* after the war. Nogales, Mexico, which was a sister city of Nogales, Arizona, was promoting a bullfight in honor of its sister city. The publicity person had asked me for an article on the bullfight. I had never seen a bullfight and was suspicious of it. I said I didn't think I would like it, but I thought I would see it for myself. He said that the meat of the bulls all went to poor people, that the bull died very honorably in a ring instead of a slaughterhouse, and it was really wonderful. He was just a public relations man from the Chamber of Commerce. He had never seen a bullfight, but he dutifully read up in the library and eagerly addressed himself to writing a promotional brochure, a copy of which he enclosed in his letter.

He had done a first-rate job of answering in advance the complaints of non-aficionados. Unfortunately, there was not a word of truth in what he said. The bullfight is a sadistic spectacle remaining from the dark ages of our civilization. The bullfight that I saw — oh God — was unbelievable. There were six fights. Each matador had two bulls. There wasn't one with even a remote contest to it. I took one of the cushions and fired it at one of the matadors. It was a bit of a rainy day, and I got him in the neck with the soggy cushion as he went by.

I later did a program for *60 Minutes* about bullfighting. These are the facts. In general, the horse has its vocal cords cut. He is blindfolded so he can't see the bull coming after him. The bull can get under the protective covering and injure the horse, and in the old days they didn't put any protective covering on them at all. The purist still likes it that way. Once the picador has gotten that pick through the neck muscles, all the bull can do is move his horns up and down. As a result, when you see all this fancy dance with the matador, there is really nothing the bull can do about it. When you see a bloodless fight, where they are not allowed to use picadors, then the bull can go sideways and can get the matador. But with modern medicine and penicillin, there is very little danger to the matador. What you are watching is an animal looking for a place to die. He is looking around while the people are screaming. Over and over again they thrust that sword, maybe five or six times, and he is sometimes still not dead. When the matador does the final sword, it is supposed to be right over the horns into the neck, through the lungs, and into the heart. But if the sword just pierces the lung, blood pours out of the bull's mouth, and he can't die and be over with his misery. Then they bring over the cloth and try and make him come around again — to stab him again. It is just the most gruesome sight. I went back to the Tucson Public Library afterward, and there were eighteen books about the bullfight. Not one of them was critical, and they were written by some of the best writers in this country — Hemingway, Michener, and Mailer. Michener and Mailer have gotten to be friends of mine, but we have never agreed about the bullfights.

When I started writing and talking about bullfighting, I soon found that you just can't limit yourself to one area. So bullfighting became the starting point of the Fund for Animals. I decided to "put cleats on the little old ladies in tennis shoes." I didn't mean any putdown of the sex, age, or footwear of these people, but the battles had been fought on a very unproductive level. I mean, they wrote poems about animals and went to little meetings and church groups. Right from the beginning I wanted something different. We weren't that big to start with, nor did we have that much money, but it sure grew rapidly. We have been involved with a large variety of causes ever since.

We were told that it would be absolute death for our society to go after 113 hunting. But we had always been a high-risk society, and I was so sick of the hunters. My book *Man Kind* was the first effort, and then *The Guns of Autumn* was made from it. Those two projects were the first major attacks on sport-hunting. Not only hasn't it ever hurt us, it has brought us recognition from people who realized we were willing to take on such a powerful group. The only way it is really going to change is when you get a governor elected who has the courage to say, "I am going to put people on both sides of the hunting question here and see what is really needed."

The hunters always claim that they don't want to kill and that they love communing with nature, but that is just nonsense. If they really didn't want to kill they wouldn't kill. They go out there and stalk animals. One fellow talked about the oneness of nature and how he and the animal just stood there, and I said, "Sure it is a oneness of nature. There is going to be only one left when you get through!"

I think that sealing is one of the most frustrating causes to me because for so long we made so little distance with it. I have seen it for so many years and seen that same Godawful clubbing. It is the worst because it is the killing of a baby animal next to its mother. It occurs on the ice, where she is power-less to defend it. Any other place she could defend it; but on the ice where it is born, she is helpless. It has to be born out of the water, and it can't swim. She has to stay by it and see it clubbed to death and skinned in front of her eyes. They hate it. They howl. They sit with the baby's skinned body. They make these pathetic sounds. They try to get in front of the babies to save them, and of course the sealers hit them every now and then. We have seen mother seals by the carcasses, and they had been there for days. I think it is the rottenest thing that man does to an animal. And for nothing, not even for a fur coat. It is just for the lining of a glove.

But thank God in 1983 the clubbing was stopped — not only by our efforts, but by the work of a lot of other people. However, though they will no longer club babies on the ice, they nevertheless announced plans to shoot the seals on the ice. In its way, this could be worse.

Sealing is about the worst thing I have ever seen, but cruelty is a strange thing. It is cruel, too, when a kid comes home and his dog has waited all day for him, and he goes by him to his room and just shuts the door. That is terribly cruel too. That animal's whole life is that kid, and he might be better off to hit it with something than do that. It is terrible.

Another issue involves the arrogance of scientific researchers who feel they have the right to do anything they want with an animal. We are in-volved in a case that you almost have to see to believe. It is a case of cruelty to monkeys in a laboratory. It is the first time that a man has been in criminal court for a cruelty-to-animal statute. A lot of states exempt laboratories, but we brought the case in a state where they don't. These particular thirty-two macaque monkeys were experimented on for eleven years. I think he is down to sixteen now; one more just died. They take a nerve out of their arm to make their limb dead. Then they try to teach them to use the limb after it is dead. Food is literally thrown into the cages, and gets down in the bottom among the feces and the urine. He never had bowls for them. It was a filthy place, and his argument was that monkeys are always dirty and that anybody who knows monkeys knows how dirty they are. One guy described it as hell in there. But you can't get them on the experiments. You can only get them on the treatment of the animals.

I think the removal of the burros from the Grand Canyon is the most important thing we have ever done in the sense of the effect it is going to have on the whole wild horse and wild burro problem. The burros had to be removed because they were eroding the canyon walls. The officials had in their minds that the only answer was to shoot them, and I think we proved by doing it in the most difficult location imaginable, the canyon, that it was possible to do a successful rescue almost anywhere. They were absolutely convinced we were going to fail. Never from the moment we started did I agree. I thought a lot of times that it might take longer and I knew it would cost more, but that was based on an estimate of three hundred animals. We got five hundred and seventy-seven out. You have to remember that the Grand Canyon is one and a half times as large as the state of Rhode Island. It is a huge place.

The hardest single job in the beginning was just finding the burros. We spent $250,000 on helicopters alone. There was no way to do it without helicopters because you couldn't walk them out of there. You needed really wonderful pilots, and thank God we had them. They were marvelous. They never hurt one burro. They flew five hundred and seventy-seven burros up on a sling in the helicopter and never hurt a one. Never skinned one, never broke a back. There was never even a wound on one, they did it so gently. They all started with a kind of skepticism, thinking we were sort of nuts and do-gooders. But they came to see that burros are incredible, interesting, and bright animals.

The Central Park carriage horse bill has been a great personal satisfaction to me. The horses live in a terrible stable. One day I was driving along Fifty-ninth Street and this carriage driver was trying to turn his horse around in the traffic. The horse couldn't see where to go. All the horse saw was all these cars coming, and he stopped, kind of quivered, and he didn't really rear, but he just kind of put his front feet out, like "Where do we go?" The driver just hauled off and kicked him in the groin. I got out of the car and grabbed him by the throat, and said, "You miserable son of a bitch. This horse looks after you for fourteen to twenty hours a day, in snow and summer heat, and you keep him in a lousy stable. It is a firetrap and a fleabag, 115 and then he gets up and earns you a living." I had him right by the Adam's apple. I just completely lost my temper. I said, "We are going to have an agent follow you all summer long, and I hope to God you get a fine, and it will probably be a lousy little fine. But if you mistreat it again, if you touch your horse in anything but a decent way, we will get you. I personally will come after you."

It is quite amazing, the effect that the bill has had. It covers all horses in New York, riding horses as well. What I think is important about it is that these horses are a very public thing. Some of the kids on the streets of New York haven't had any background in animals. It's not right for them to see somebody kicking a horse or having a horse running full tilt down Seventh Avenue on a slippery street. So what if somebody says Amory cares too much about the highly publicized things. I don't necessarily, but we do an awful lot of these more publicized issues because of the humane education involved. The horses now have more decent working hours and mandatory veterinary care, and the bill involves about two thousand horses, a hundred of which are carriage horses.

The Central Park Zoo issue has been a terrible failure, but we are ultimately going to win. It took us so Goddamn long to get anything done about that zoo. I think zoos in general are amoral to begin with. You should accept that even if you are a good zookeeper and run a good zoo. You should realize that you do not have the right to imprison an animal for the mere crime of being what he is so you can exhibit him. For me, it is an issue like vivisection. In my heart I am an antivivisectionist, but in my head I know it is going on, so I want it regulated. Therefore, since zoos exist, certain things must be done. You must protect this animal from the public. You must give it decent food. You must give it decent company. You must give it as much space as you possibly can. If you do these things, then at least you are trying to run a good zoo, not just trying to make money out of the popcorn concession or the public coming through.

Zoos are a form of slavery, and I think that someday our treatment of animals will be regarded on a par with slavery, but I think that the wildlife safari parks have the idea. They let the animals loose, and they put the people in cars and drive them through. I think a monorail is an even better way. It is cleaner, quieter, and doesn't bother the animals. Central Park is an anathema to me, with its old-fashioned cages and each animal alone.

I follow reports on the animals of other countries. Turkey is a terrible place for animals. We had one of our agents over there, and she is going to bring in a report of what she thinks we could do. For all the poaching and the horrors of Africa, there are a hell of a lot of African people risking their lives. It would bring me the most satisfaction to see the kind of progress I feel we have made in this country duplicated in South America and eventually the Middle East and the Far East. It is a hard, hard fight. In my lifetime I would like to see no animals bought or sold, just adopted in shelters as long as there is a stray animal problem.

I saw a movie once that made the best impression on me. It is all about putting yourself in the position of the animal. It was made by the Pasadena Humane Society about a cat trying to cross a highway and was presented through the cat's eyes. They put the camera on a stray cat and they went where it went during the daytime. All of a sudden it is trying to cross a Los Angeles highway. You look and think, "There is no way it is going to be able to do this." You see these trucks and these terrible lights, and it is all at the height of this animal. I defy anyone to see that and not have empathy. We are animals too. Let's face it, and get the guts to be unselfish.

The Life and Death of a Beloved Ferret

Wendy Winsted, a thirty-five-year-old third-year medical student at the University of Cincinnati, has spent many years popularizing the European ferret as a pet in the United States. These small carnivores are related to weasels, minks, ermines, polecats, badgers, martens, sables, skunks, and otters.

Wendy has published articles and a book on ferrets, is currently working on a second, more comprehensive text, and has developed an operation for de-scenting ferrets. For the past decade, she has almost always had at least one ferret with her at all times. When she leaves her apartment, she often carries one or two in a large handbag with special compartments.

WENDY WINSTED: When I was a kid in Oklahoma, we had a dog and a cat. Then my mother died when I was twelve, and my father remarried when I was fifteen, and my stepmother didn't allow any animals. I guess the first thing I did when I got out of their house was to fill mine up with animals. I have a brother and a sister who like animals, but they aren't as involved as I am.

About seven years ago, a girlfriend and I had skunks. My skunk died in the winter, and you can only buy them in the spring. My friend told me about ferrets, and I decided I would try one. They were domesticated by the Egyptians before the cat. They are like Yorkshire terriers, meaning there is no such thing as a "ferret" that lives in the wild. They come from an animal called a polecat, which is a little bit bigger and very vicious. I got my first ones very, very young because I wanted to bottle-feed them. They couldn't   117 even walk, so I wasn't sure whether I was going to like them. It was about five weeks before they turned into little animals and then I loved them. I bought them from a ferret farm, where about five or six thousand a year are bred as lab animals.

For three or four years, I was one of two customers buying them for pets. All the other people would just buy them for research. I would go down and see these five thousand baby ferrets, and I would know that they were going to be used for research and whichever one I picked and found a home for or sold was going to be saved. It was really strange to go and see these thousands of little faces looking up at me and know that I could only take eight. So I would always end up taking more than I should have.

Once I went up to the farm with my ferrets Melinda and McGuinn on their leashes, each wearing a collar with bells, and had them do tricks. The people there were handling the ferrets with thick leather gloves. They called all the staff into the conference room to show them that ferrets could do tricks. They just couldn't believe they were so gentle and responsive. They finally changed their whole breeding program because I would order some and say, "Send me ones that won't bite when you stick your hand in." Now they stick their hand in the mother's nest and handle the babies. If she bites at them, they don't breed her again. They decided it was good to have stock that didn't bite anybody, no matter what. They also started breeding different colors for pets — some with white feet, dark coats, and black noses. And then they started breeding them for shorter, cuter faces. But it breaks my heart that most are used for labs, because they're such trusting, loving little animals.

To be kept as pets, they have to be de-scented. I developed a surgical technique, and I have taught a lot of vets how to do it. When they are five or six weeks old, you make an incision and dissect down until you find the scent gland and you remove it. It is a lot like the surgery for de-scenting a skunk, only a ferret has more connective tissue and the gland is harder to get at.

The two ferrets I have been closest to were Melinda and McGuinn. I favored them over everything. As a matter of fact, now I am not too crazy about my cat and I feel guilty about it. The cat is more obtrusive. When I'm reading, she will sit right on the book. Ferrets are more independent. By the time they decide to sit on your lap, you are so flattered that you enjoy it.

They went with me everywhere. I have a handbag with a little litter pan and a little bed. They rode in there, and they slept on my lap in class. They knew in the subway I wouldn't let them get down to walk around. Sometimes people would hold out their hand and they would crawl up on it. The only time we ever really had severe arguments was when somebody with a shopping bag sat next to me. Then I had to fight Melinda because she just

had to go into the shopping bag. Sometimes people would let her, and sometimes they wouldn't. If I wouldn't let her, she would finally roll over on my lap. She knew that trick usually would get her about anything she wanted.

They are very good-natured and adaptable, intelligent and fearless. While skunks or cats are terrified of the vacuum cleaner, the ferrets like to chase it. They think nothing is going to hurt them and everything is made for them to have fun with. Most of them who have never seen a dog will run right up to the dog and want to play from the time they can walk. They're very easygoing. I keep thinking I should learn to be more like they are, just accept things as they come along. And I don't think I've seen an animal that likes to tease so much as they do. They will run out and nip your toes and run away, wanting you to chase them.

When Melinda had her last litter, she got a mammary gland infection and I had to take her kits away from her, so she adopted me. She kept washing me like her baby. Finally she got better and I gave the three-week-old babies back to her, but she woke me up in the middle of the night. There were her six babies right around my neck, and she was licking me. I was still her baby, but she couldn't move me to the rest, so she brought everybody together.

The ferrets have enriched my life greatly. People come and go, but through it all there is my little family that is always there and consistent. They never betray me, run off and leave me, or lie to me. And my best friends are people I met with the ferrets. They are people who have basically the same kind of values that I do. They're generally sensitive and caring.

I've gone out with a lot of veterinarians. Then there have been guys who have said it's either me or the ferrets and I've said goodbye. I've never been married. I'm sure that's not because of the animals but because of my own hangups. I just am not able to make commitments.

I sometimes feel pathological that I am so attached to Melinda. The death of McGuinn really depressed me a lot. It was more than two years ago, and I still miss her. She was with me for six years almost twenty-four hours a day. I spent more time with her than with any person. I probably was almost as upset at her death as I would have been at my father's, whom I haven't seen in five years and who doesn't particularly get along with me. There was a certain rapport McGuinn and I had. I could tell how she was feeling by watching her reactions to things. Actually, I'm glad that it was a sudden thing rather than her being sick. I would have been worried crazy until she died. I would have just gone nuts.

What happened was, when I came home one evening, Melinda came out and I didn't see McGuinn. When I started to go to sleep, I thought, "I should get up and go look for McGuinn because she might be in trouble somewhere." So I got up and looked for her. Then I remembered that morning I had felt her underneath the mattress, sort of down by my feet, pushing the mattress up with her head the way they do when they're under things. I remembered thinking, "I wonder what McGuinn is doing under the mattress?" So I looked under the mattress, about a foot up from the end of the bed, and I found her. Dead. I think she went to sleep and suffocated. I do know that ferrets suffocate easily. I guess I couldn't believe it for a while. I took her out and sat there looking at her. I kept thinking, "I don't know what to do. I don't know what to do." I was just in a state of shock. I wasn't aware of how Melinda was reacting. I basically sat there and said "Oh, McGuinn" over and over. It seemed odd that I wasn't crying. I expected to be hysterical and flip out, to just completely fall apart. I thought, "Well, what is wrong? You know, you are not falling apart, and you aren't crying." I must have sat there for four minutes or five — maybe it wasn't that long; it seemed forever — saying "Oh, McGuinn." I remember thinking maybe somehow there was a way to go back and undo this. And then I thought, "Well, I've got to call somebody." Then I realized, "Well, who are you going to call?" I felt there was no one I could call who would understand what a big loss it was to me. If your mother dies, that's okay; everybody can relate to that. But if somebody's ferret dies, it's supposed to be different — just an animal, you know. Why can't people see that it's the same? I didn't think people would so much think that I was crazy as that they wouldn't be sympathetic.

I'm a very intense person. I'm not sure it's a blessing, really. I feel as if I have spent my life hitting the ceiling and crashing on the floor alternatively. I guess that is why I feel that many people wouldn't understand how I see the world. For example, ferrets get crazy when they're playing. They don't look where they are going; they just jump around, and if they bang their head on something, they jump in another direction. And when I look at that, I would never mistake it for anything other than playing. Yet some people think they are having convulsions or something. To me it looks like a little, happy ferret, jumping around and leaping. To me it looks like joy.

The Pig Lady of Ramona

It took me a year to find out if Judy Van der Veer really existed. I learned about a possibly mythical "Pig Lady of Ramona" from some youngsters at a livestock fair. They had heard that somewhere in the remote hills an old lady shared her house with grown pigs. Finally, after following a chain of leads I located her, her current pig, 121

Little Brother, and her extended family of farm animals, including a white donkey, several cattle, and a horse.

Judy Van der Veer is a rancher and a writer of children's books. She wrote about Wallace, a pig in her past and the hero of his own book: "It would be Wallace's mission in life to prove to the world that a pig is a clever and charming animal."

JUDY VAN DER VEER: I was born in 1907 in Oil City, Pennsylvania. The first thing I can remember is being two years old and sitting on a Shetland pony named Girlie, who lived next door. I was madly in love with her. When nobody was looking, I'd crawl under the fence and go to play with her. As soon as I would be discovered, there would be screams of wrath and a spanking. But the next time I got the chance, I was over there again.

My father owned a jewelry store. But my mother, who loved animals, was such a baby that she wanted to go back to San Diego, where her parents and sisters lived, so we moved when I was eight. That was too bad for my father, and he got a horrible job being an inspector in some kind of factory. He was a dreamer. He should have been an inventor. But he was cold in a way. I had a girlfriend at school, and her father had a drugstore and we would stop there on our way home from school to get some goodies, and she'd rush up to her father and say, "Oh honey," and he'd say, "Oh honey, how was school today?" He'd hug her and she'd kiss him. It looked so nice that the next time I saw my father I called him honey, and he said, "Don't call me honey!"

I hated school, and I am a high school dropout. It was the best thing I could have done. I got a job on a dairy ranch taking the cattle to graze. It was dull, but I liked it. I got paid $1 a day for about sixteen hours. Of course my parents had fits, but nobody could make me go to school anymore. Actually, I taught myself to write well by not going to school and having the chance to read good books. My first book came out when I was twenty-eight, *The River Pasture.*

I never married. I was having so much fun all the time, going to dances and riding with a bunch of guys. And then my little girl chums would be married. When I saw them, they'd be holding one baby and then there'd be three or four more little kids. They'd look so wistfully at me. I think I've had more fun than anybody.

I feel as though I was born knowing how I felt about animals. I remember when I was a little kid in grammar school, the teacher was explaining in class about meat. She had visited a slaughterhouse of her own free will and enjoyed it. Imagine that. They had a stadium, and you sat and watched as they strung them up and cut their throats. "Doesn't that hurt?" said a little kid. "Oh yes, but that doesn't matter," I remember her saying. I knew she

was wrong.

As a young girl, I had pet lambs. I had one that would go squirrel hunting with the dogs because she was very playful, and when the dogs would run, she would run with them. Our house at that time had an open deck instead of a porch, and on some summer nights the little lambie would climb the stairs and she would dance. You would hear her little feet, and she would buck-jump and caper all night long.

The house I live in now belonged to a pioneer family. It's about a hundred years old. When I bought it, it had 560 acres, but I couldn't pay the taxes on it so I split it. I only have 240 acres now. I didn't like those pioneers. They were very hardhearted. They got as far as Texas and came the rest of the way in an ox-driven covered wagon. When they arrived here, they took the lead ox that had nearly died dragging them across the desert, mountains, and passes and traded him for wood to build their log cabin. You would think they would have let the poor beast retire. Eventually the old log cabin was torn down and my house was built. It was a schoolhouse at one time before I got it.

I live with six dogs, two horses, a third horse that's not mine, one burro, nine cows, the biggest steer you ever laid your eyes on, and the pig. Mamie is the oldest cow, maybe around twenty. She gets fed special. If the others come first, they will eat her food. She was born in the corral when a bunch of us were sitting on the front porch. Calfalier is the beautiful steer. He reminds me of the legend of Europa, the white bull who enticed a maiden to get on his back and carried her off. One of his grandfathers was a fancy Charlay named Cavalier de Paris, so I named him Calfalier. His mother, who was part Hereford, lived with me. My cows arrange their own marriages. I don't have a bull because I don't want any calves, but she found a gentleman and had Calfalier. She died when he was born in December 1969. He used to sit on my lap and watch television when he was a baby.

Once he escaped over the fence and got into the cattle ranch next door. I couldn't ride out and get him alone because it's too hard for me to drive him away from a herd by myself. Finally, the rancher got him in the catchpen and chute to bring him back by truck. But before that he phoned and said he was going to ship a lot for slaughter and he would be glad to take my steer for free. I said, "Thanks, but I want him back." He said he was worth close to $1000 what with the price of beef. There was no use explaining it all to him. The only money I would make from Calfalier was by writing about him, not killing him. But most people think killing is all a pig or steer is good for.

I always loved pigs. I got my first pig when I was a girl. I was on a horse, driving cattle to a well, and I saw this strange animal from a distance. The cows skittered and my horse shied. I got a little nearer, and there was a young pig happily eating acorns. She followed me back to the barn and I named her Pig. She was brown.

Over the years I have raised several little baby runts — the ones they just hit over the head and throw into dumps. When they are little, I keep them in the house. When they are too big for the house, they live on the front porch. When they are too big for the porch, they live outside the back door, but they always come around to let me know it's time for their bottle. They squeal, come up on the porch, and look in the door. I remember once I had some visitors, a bunch of ladies all dressed up, and my pig opened the door and walked in. They screamed. My pigs were sort of housebroken. I have heard that people live with great big boars that lie down with the dogs by the fireplace. But I found that when I got a pig to go on paper, he would get so pleased with himself 'cause I praised him that he would pick up the wet paper and go all over the house, shaking it. He was celebrating. Baby pigs are very playful, like puppies. They pick up a stick and run around, and the others chase them. They have fierce little teeth and bite each other. I could always walk my baby pigs on a leash. They would trot along like a puppy dog.

When they were old, they died. Or they would finally get down and couldn't get up. Those I had to have shot. You know, a pig is very incorrectly built. They are too heavy for their little feet and legs. Wild pigs are probably much leaner. I try to keep their weight down, but when they finally can't stand up they really are miserable. Arthritis sets in. But I have kept all of mine till old age. I read a book about pigs, and it said that the life span of the domestic pig is not known because no pig lives to die of old age. Except mine, that is! They live to sixteen or seventeen years. At one time I had as many as three or four. The neighbors couldn't understand it at all. Why would anyone have a pig and not eat it? A pig being slaughtered for food won't happen to any pig I know.

I have heard that pigs are supposed to be very smart, and I can believe that. They know their names, and they certainly are smart about learning how to open doors and how to con you. I had one little pig who loved to go to the barn because while I was doing chores, he would go to the feed room and have a great time. So at either the morning or the afternoon time for going to the barn, he would come to the house and start squealing for me. He would watch me and then turn around and wait for me to follow. He had it all figured out. After I realized how smart they were, I decided to teach one little two-month-old pig a few tricks, like you do a puppy. So, on command, he learned to lie down, and I would rub his belly as a reward. Then he would shake hands. After a while, though, when I said, "Come on, let's do it," he would take his nose and press it on the floor and brace his little legs. I couldn't pry him loose. He was through doing tricks, and he never did them again.

I liked them all, but I wrote a book about Wallace. There was a big 125 commercial pig ranch next door (where there is now a thoroughbred horse ranch). He was one of a litter of pigs to be fattened and slaughtered. I don't know how he knew it was a bad place, but he walked right out of there and came a mile to my place. He followed me over to the barn, so I picked him up and put him in the house. Pretty soon guys came along and said, "Have you seen a pig?" I said, "What pig?" So he escaped a bad fate. When he grew up, he liked to sit on the back porch with me in the cool of the evening. It was as if the pig and I had our arms around each other's backs. We would watch these trucks go by just jam-packed with pigs on their way to market. He didn't know how lucky he was, but I would tell him, "You might have been there too." It was fun to write his book, *Wallace the Wandering Pig.* He was spotted, and they had a picture that looked just like him on the cover.

Now I am down to one pig. I got the pig in the sixties from some flower children, who had named him Little Brother. They kept him with a chain around his neck. Then they moved out, leaving baby chickens in pens and starving dogs. He was used to people and didn't have a reason to think people were dangerous. I guess he should have. No one tries to steal him. He wanders around the road and sometimes takes sunbaths. One time somebody drove past my barn, then backed up and said, "Do you know there is a dead pig by the side of the road?" I said, "No, that pig is just taking a sunbath." A pig is inclined to be indolent and lazy. Little Brother sleeps most of the day, in summer especially. They like to sleep in the shade or cool mud. He loves to be scratched and petted. The bristles over his body are stiff, but behind his ears he is smooth as silk. They make toothbrushes and other things from their bristles. He can make very loving sounds. He goes "ooh, ooh" like an Indian when he wants to be petted, and when he is mad he makes a very shrill, annoyed sound.

The only time pigs are not friendly is when they are eating because they think you are going to eat their food. They get greedy. You have heard "hungry as a pig" or "eats like a hog" — and they really do. Pure enjoyment. Little Brother eats table scraps and just about anything else. I have no problem with garbage — just toss it out the back door. He comes over and hangs around, waiting. I try not to feed him too much. I don't want him to get too fat because it will be harder on him now that he is old. Well, he has gotten stiff. When he lies down too long, he limps when he first stands up.

A thing that's odd to me is people who say, "I like dogs, but I hate cats," and I have a neighbor who hates coyotes and will shoot them, but he loves bobcats. I think we are in the process of building ourselves a Kingdom Come. I am a violent pacifist. I'm against war because they might destroy one tree or one rabbit. That is enough to turn me against it.

I broke my leg in January of 1981. I was riding a colt that was spoiled and green, and I was moving the cattle. Somebody's dog spooked them, and they ran up behind my horse. He took off on a dead run so quick that I lost my balance completely. I felt like such a fool. I have not been able to walk right since then. And I have had three cancers and had a mastectomy some years ago. You are supposed to have traumatic feelings about things like that, but I never did because I figured I would rather have my life than my breast, and I wasn't using it for anything interesting then, anyway. But the cancer really started acting up six months ago. I haven't given up yet, but they don't give me much hope. They just say the best they can do for me is to keep me as comfortable as possible. Since I don't know what will become of the animals, I just have to keep on living. Sometimes, when the pain gets too much, I think I could take a few more sleeping pills and get out of this whole mess. But then I remember my responsibility.

Bat Conservation International

Merlin Tuttle, curator of mammals at the Milwaukee Public Museum, is a world authority on bats, and his high-speed photographs of them rank him among the world's great wildlife photographers. His pictures and articles have appeared in *National Geographic* and many other scientific and popular publications, and his current research on frog-eating bats is featured in a BBC film that has won several top awards.

He is divorced, has no children, and lives alone on the sixteenth floor of a condominium a mile and a half from his office. He has only a few photographs of bats at home, but in his sixth-floor office at the museum, the walls are covered with his poster-sized, lushly colored pictures.

MERLIN TUTTLE: I distinctly remember certain events that occurred when I was less than two years old, and most involve interaction with animals — a toad, a baby bird, and monarch butterfly caterpillars. Nearly all of my fondest memories involve animals and the outdoors.

I was born in Honolulu, Hawaii, forty-one years ago. My father was a biology teacher, but he was not particularly interested in mammals. He preferred to watch birds and was also interested in keeping reptiles and amphibians as pets. I was kind of a strange character in that by the time I was nine years old, I already had decided to be a mammalogist.

By the time I was in high school, my family had moved to Tennessee, where we lived near a bat cave. I soon noticed that the bats were present only in spring and fall, suggesting that the cave was used by migrating bats. I identified them as gray bats only to discover that all the mammal books at that time agreed that this species was nonmigratory.

127

While a senior in high school, I convinced my parents to take me to the Smithsonian Institution in Washington, D.C., to discuss my findings on possible gray bat migration with their bat specialists. I brought my three years of observations suggesting that gray bats were migratory, and they gave me several thousand bat bands, saying, "Here's your chance to prove it." My whole family became involved, and my father ended up spending hundreds of hours helping me capture and band bats.

Only a few months after we began banding gray bats, we heard from local old-timers about a cave where bats hibernated. Since it was about a hundred miles north of our home, we doubted its importance to us. We assumed our bats would go south for the winter. Even so, curiosity got the best of us, and by an extreme stroke of luck, we found our banded bats in that cave! They had not only migrated, but they also had gone north instead of south.

My doctoral dissertation on population ecology and the behavior of gray bats was eventually completed at the University of Kansas. It soon became obvious to me that bat colonies were declining rapidly. Even the ones that I once thought had the best chance of long-term survival were declining. Six years after I finished my dissertation, when I went back to recensus, I found a 54 percent decline in bat numbers, even though the census was limited to the groups that I had thought were least likely to decline.

In one case, I returned with friends to show them the spectacular emergence of two hundred and fifty thousand gray bats from Hambrick Cave in Alabama. At almost the same time every evening, you could see this big, dark column of bats, 60 feet wide and 30 feet high, going all the way to the horizon. The sound of it was like a whitewater river. Such emergences of bats are among the most spectacular sights in nature. We were all excited, with cameras ready, but the bats never came out. It was quite a shock when it dawned on us that the bats were gone. I had had a lot of emotional attachment to them. These bats had played a major role in my doctoral research. We went into the cave and found sticks, stones, rifle cartridges, and fireworks wrappers beneath the former bat roost. And by subsequent winters I knew they had been killed. Many were banded, but they didn't show up at their traditional hibernating caves. There were so many ways they could have died. Even a single blast from a cherry bomb could have severely damaged their extremely sensitive hearing, making it impossible for them to use their sonar. Hambrick Cave was 5 miles from the nearest human habitation, and you could get there only by boat. It was one of the last places in the world where I expected bats to be destroyed.

A big problem for bat survival is that bats are, for their size, the slowest-reproducing mammals on earth. Most bats rear only one young per year. In
contrast, if you took a pair of meadow mice and gave them everything they

130

needed for survival, theoretically they and their progeny could leave a million meadow mice by the year's end. If you provided an average pair of bats with the same opportunity, in one year there would be a total of three bats — mother, father, and baby.

Bats are so sophisticated and highly specialized that their life strategy demands long survivorship. A meadow mouse is basically just an eating and breeding machine. In contrast, a young bat has to learn to use very sophisticated sonar and hunting techniques, to fly, and to find sometimes distant feeding grounds and follow migration routes to caves hundreds of miles away.

In north temperate zones, hibernating bats must survive for as long as six months without feeding. Survival depends on not running out of stored fat reserves. The mere passage of a person through their cave disturbs hibernating bats enough to cause them to burn up from ten to thirty days of fat supply. If people repeatedly enter bat hibernation caves in winter, entire populations can exhaust their fat reserves and die.

There are well-documented cases in which thousands of bats have been killed by people who went in and raked them off the walls with sticks. The bats ended up being stomped on or drowned in pools. They're completely helpless when they are hibernating. Their hearts are barely beating, and their body temperature may be hovering just above freezing. It might take them as much as an hour to arouse and be able to fly to escape.

When they come out and have to migrate again, they face additional problems, because people have destroyed their stopover caves. Lacking these, they must push themselves to exhaustion in longer travels per night, or they may have to roost in places where predators can find them. When they get back to their summer quarters, their problems continue, because when they rear young they are still extremely vulnerable. The young can't fly for close to three weeks, and during that period people entering the cave can quite inadvertently cause thousands of young to die when panicked mothers either drop them while attempting to escape or move them to less suitable locations. The ceilings of caves used by bats are often etched by the bats' claws. This makes them suitable for maternity purposes, where the young need a rougher surface on which to cling. The etching is also a great advantage in insulating young bats from the cold — it's like a foam mat on a concrete floor. In some caves there are notches up to an inch deep. If bats are forced to move to new caves, conditions are less ideal, and survival is reduced.

I had kind of assumed some caves would be safe indefinitely. But it's just not true. Bats are so vulnerable that it's rarely possible to say that any colony that isn't entirely protected is at all safe. There are still some great bat colonies on most continents, but they are disappearing rapidly. In Latin 131

America, millions of highly beneficial bats have been killed because of ill-advised vampire bat control programs. The people involved often can't identify one bat from another. Tens of thousands of caves and other roosts have been destroyed, simultaneously killing countless other animals. In the United States and Europe, amateur cave exploring and pesticides are big problems. In the Pacific Islands and Asia, and to a lesser extent in Africa, the introduction of firearms has caused declines. Too many bats are shot for food. Nearly everywhere, bats are threatened with habitat loss and pollution as well. On top of all this, countless thousands are killed simply because people fear them.

But it isn't surprising to me that bats rank high on many people's list of most-feared animals. Bats are nocturnal and secretive, and people fear most the animals they understand least. There are nearly a thousand species of bats in the world, roughly a fourth of all the world's mammal species, and most are harmless and highly beneficial. Historically, only Westerners by and large have been afraid of bats. I suspect this is probably because bats in Europe and the New World are mostly small and difficult to observe. In places like Asia and Africa, many bats are large and conspicuous. People do not fear such bats but simply eat too many. In Asia, bats are held in great esteem and are used as symbols of good luck and happiness. Chinese art is full of stylized bats.

In the United States, the image of bats took a horrible turn in the early sixties because of certain hypotheses regarding rabies. Several experiments appeared to show that bats were unique among mammals in that they could carry rabies without themselves being harmed. This led to the hypothesis that bats served as reservoirs of rabies for wildlife. News media headlines from coast to coast and around the world greatly increased nearly everyone's fear of bats. Subsequently, researchers set out to test their hypotheses. First, they did *not* find more rabies in wildlife in areas where bats were most abundant. Then they reevaluated the original tests. Brain tissue had been injected into mice, and the mice had died of rabies-like symptoms. But the virus involved turned out to be one that is harmless to bats and people, though lethal to mice. So the original bats weren't even rabid, and the theory was wrong. But a massive destruction of bats had followed. It was quite appalling. There is now an attempt to reeducate health officials and the public. All mammals can become rabid, but less than half of 1 percent of bats do, and even these rarely become aggressive. I have not seen a single aggressive bat in more than twenty years of studying them, though I have handled hundreds of species around the world. I simply warn people not to handle any sick animal and to assume that any bat that permits itself to be picked up is sick.

Worldwide, bats are sadly misunderstood. Most are harmless and highly beneficial. Bats are the only major predators of nocturnal flying insects. Individuals can catch hundreds of insects per hour, and some of the largest colonies catch more than half a million pounds of insects in a single night! In tropical areas they are even more important. Bats often constitute more than half of all mammal species in rain forests, and their pollination and seed dispersal activities are critical to these forests' survival.

Bats are among the world's most diverse and fascinating mammals. They live nearly everywhere. They range in size from the world's smallest mammal, a species weighing nearly a third less than a penny, to giants with 6-foot wing spans. Some are bright yellow, others snow white with long angora-like fur. They are highly intelligent and social and may live up to thirty years or more.

Bats are the only flying mammals, and their acrobatic maneuvers are unparalleled by any bird or manmade device. Most bats use ultrasonic signals for navigation and communication. They have mastered the sky just like dolphins have mastered the sea. They can send out pulses of sound that, with extreme precision, can perceive motion, distance, speed, trajectory, and shape. So it appears that they can "see" all but color with sound. They can detect and avoid obstacles no thicker than a human hair, and millions sometimes fly at once in a large cave without jamming each other's sonar. Their abilities far surpass our present understanding.

When I first started training bats for photographing them, I had no idea that they were as intelligent as they are. Time after time I use the same approach in trying to get a bat to do something as one would do with a human being. First I establish trust. Individuals vary greatly. That's why I usually catch several and work with them for an evening, releasing those least likely to be trainable.

I have studied bats for most of my life, but I am continually amazed at their many sophistications. I especially remember one bat that I released back into the jungle only twenty-four hours after its original capture. It already had learned to come on call for food. About fifteen minutes after its release, it apparently followed my voice to find me back at my cabin, where it repeatedly attempted to land on my hands to be fed. This bat had lost most of one ear in a prior accident and was easily recognizable. When I found him a month later, roosting in a large hollow tree some 2 miles away, he still must have recognized me because he showed no fear.

In Thailand, I trained a small species of flying fox bats to come on call within only a few hours. They were extremely gentle, and I often held and petted them. In turn, they would snuggle into my hand and lick my fingers. They were such super-attractive animals that I really hated to turn them loose, especially knowing that they likely would end up in a Thai restaurant.

A botanist friend of mine is interested in flying fox bats because they pollinate the flowers he studies. They have 6-foot wing spans and are active by day, soaring up to a thousand feet above the jungle looking for flowers.

They have declined rapidly due to overharvesting for human food. Thus, they naturally must be extremely wary of man. They roost in pairs, possibly mating for life. In one instance, my colleague sneaked up on a sleeping pair. The male woke up, spotted the apparent danger, and quickly flew away. However, the female remained asleep. The male returned three times, slapping her with his wings, apparently to wake her up. They finally flew away together.

Bats are born in a relatively advanced state, some more than a third the size of their mothers. Babies are very inquisitive and have behaviors we attribute to other intelligent animals. Each mother is very solicitous about protecting her own baby. While roosting, the upside-down mother holds her young enclosed in her wing and against her pectoral breasts. In some species, parents take turns babysitting while one hunts for food.

In 1979 I participated in a bat symposium in England. Afterward, my good friend and ally in bat conservation, Dr. Robert Stebbings, invited me on a week-long tour of his study sites in England. We discussed the world-wide problems of bat conservation, how really serious the declines were, and how difficult it was to get people interested in helping. So we decided to found Bat Conservation International. We are now launching a fund-raising campaign, using a brochure titled "Why Save Bats?" So far we have had very positive responses from people who have seen it. Many conservation organizations assumed that the public all hated bats, and money couldn't be raised for bat conservation. I don't believe that at all. People simply won't fund anything until they understand it. Almost everybody is fascinated by bats. I hope we can turn this fascination into: "My God, aren't they amazing. I didn't know, but they are really neat animals."

The four largest summer bat colonies known in eastern North America would not exist today if it had not been for my calling attention to their plight and getting them protected. I obviously take considerable pride in having accomplished something like that, though I don't think that any kind of animal is necessarily more deserving of conservation than another. I just happen to be personally very fond of bats. If you work with a species like the gray bat for twenty years, you may get to see the same bats again. And it's like a reunion when you find a banded one that you haven't seen for fifteen or sixteen years.

I'm an extreme lover of nature and natural beauty, and bats play a very important role in that for me. For example, even in dating women, I find no potential for a long-term relationship with a woman who doesn't appreciate wildlife, because to me many of the most thrilling and fascinating things in life involve being outdoors with wildlife. I can't imagine sharing my life with someone who didn't share the same kind of enthusiasm.

Doctor Dolittle Lives!

> "You know, Doctor," said Jim, "when you brought me back here
> to my homeland years ago the story of how you had cured my
> toothache and got me away from that wretched circus, soon
> spread even as far as Timbucktu . . . Your monkey here, Chee-
> Chee, told me that the same thing happened when you stamped
> out that sickness which was killing off his people. You don't
> realize, John Dolittle, how widely known you are among the wild
> animals of the world."
>
> —Hugh Lofting, *Doctor Dolittle and the Secret Lake*

Doctor Dolittle could speak to animals, according to his creator Hugh Lofting. Jim
Arden, forty-six, also has spent his life speaking to and caring for all manner of
creatures, and he now seeks to inspire his young students at Greenwich Country Day
School, in Connecticut, with the same love for animals. He and his wife, June, live in
a large elegant home that was once part of a large estate and is now part of the school
campus. June has converted the carriage house into a ceramic studio, and Jim has
made the house a veritable museum for his huge collection of vintage movie posters.
Animals are integrated throughout their home and grounds. Snakes, a tortoise, a dog,
and a ferret have their places. His classroom is reached after an easy stroll along a
tailored forest path. There, additional animals are housed: a tarantula, a scorpion,
amphibians, and more snakes.

JIM ARDEN: When I was growing up, the only white people I knew were
people I met in schools. The place I lived was called Haw Creek, a little
village where most of the poor black people lived. This was in Orangeburg,
South Carolina, where I was born.

Because my mother and aunt were afraid of snakes and spiders, I guess
I picked up the habit of being afraid for no other reason. When I was about
eight or nine years old I was at a playground, and a little girl put a snake
down my neck. She yelled "Snake!" as she did so, and the panic I had was
complete. I ran without looking in any direction; I ran into people and swings
and ended up right in the road. I got hit by the first car and was thrown
maybe 4 or 5 feet. I was badly lacerated but psychologically traumatized more
than physically. I was taken to the hospital, and I was gibbering — just
terrified. By the second day I calmed down, and my mother bought me a
book entitled *The First Book of Snakes: The Children's Book of Snakes*. She was
wise enough to realize that her fear had caused my problem. Once I got   135

the book and read about snakes, they became less frightening. So the following summer I went out and caught my first snakes and brought them home in breadboxes. My mother, who was still afraid of snakes, was happy that I had overcome my fear, but a little unhappy that it had been overcome to such an extent.

Now, many snakes other than rattlesnakes vibrate their tails. One day I brought home a black pilot snake. The local people call it a thunder-and-lightning snake. They were sure it was poisonous, and because I could handle that snake, they thought I had a special gift and was charming it. It took me a while to convince them that the snakes I had weren't venomous and that the tongue wasn't venomous.

In Haw Creek, the people did very little food shopping. They killed wild game, and I was very thin because I wouldn't eat squirrel for emotional reasons. And I wouldn't eat opossum. I wouldn't eat raccoons. I was so fascinated by those animals, and I couldn't see them alive one minute and the next converted to food. So I ate primarily hominy grits and vegetables.

My father was an American Indian from South Dakota. In the late twenties and early thirties he cut his hair very short and passed for black; in those days it was economically better to be black than Indian. He moved to Manhattan to live in Harlem, and my mother met him there and they got married. She had run away from home and was working as a domestic. I think she was only seventeen or eighteen when my father left to go out for a pack of cigarettes or a loaf of bread and just didn't come back. When the war broke out we discovered him through the Red Cross, and my mother was able to divorce him.

When I was a young boy, I didn't realize how far into wildlife I was and what a bore I was to most of my friends, who were interested in going outside and trading baseball cards or playing ball. I'd be at home with books written by Seton Thompson, Desmond Morris, and Frank Buck or *The Trails of Africa* by Teddy Roosevelt. I still have the books. In retrospect, I find that I no longer admire what many of them did, but at the time they really excited me. I still admire Desmond Morris, but the rest of them killed things and exploited animals. But these people really began my interest in wildlife.

My mother moved me and my brother from South Carolina to Stamford and Norwalk, Connecticut. At an early age, I realized I was listening to a different drummer. I flunked science and biology in school! I had a greater interest in the subjects than either the teachers or the students around me; but I was not being encouraged, and was being told continually to get some sort of manual trade because my future was there and not in education. Meanwhile, by eleven, twelve, and thirteen, I was bringing home every little wounded animal I found. When I was thirteen, my mother sent me and my brother to a Catholic boarding school. Her relatives with whom we were living died, and Mother couldn't find a place that would take children, especially a family with black children.

At boarding school the nuns thought I was very peculiar. Their orientation was that snakes were akin to the devil, so my interest was considered bizarre. But they also used it, because we had our own vegetable gardens there. It was my job to go into these areas and make sure there were no snakes. I used this opportunity to tell them that there were more snakes than I actually saw, so I became an animal expert. I was only a little boy, but they had faith in my knowledge of wildlife.

My mother's big worry about me was that being interested in animals was not going to take me anyplace. But she had a great gift. Every day when she would come home from work she would say, "What did you learn today?" She did not mean in school, she meant about life. And I would tell her why the sky is blue, or that whales were mammals and not fish. So in a way I taught her things, and at her parties I would hear her telling friends things like sharkskin is made out of teeth. I was so pleased that she was proud of me.

My brother and I didn't remain close. He was athletic and sports minded, and my interest was in books and wildlife. I also developed a great interest in the theater and read many books about acting. During that time, my mother worked for a very famous stage actress from England, and she kept telling her she had a son interested in theater. Because she was tired of hearing my mother, the actress invited me to come to her home and do a reading for her. Mind you, I had never done anything at all in theater, and she asked me to do the Mark Antony speeches from Shakespeare. I had never read it, because when I was in junior high school I was in the section that didn't read Shakespeare. Anyway, I gave her the reading cold and she liked me. She gave me a letter of introduction to a famous actor at the Stratford theater who was doing *Othello*. Then I moved to New York and went to acting school for two and a half years and supported myself singing calypso and Negro spirituals. I even sang with an Israeli folk group. Meanwhile, I would go to the Museum of Natural History and spend hours. I studied all their field guides and took every kind of free course or seminar available. I also spent a lot of time at zoos and game parks.

I had a one-room apartment in the warehouse part of S. Klein's, a large department store in Union Square, New York City. Part of the deal was to be out of the building by nine in the morning, and I couldn't be back before six. I had a few snakes in shoeboxes and a few terrariums, but it was very difficult to keep them the way I wanted. I also had a lot of trouble keeping roommates because my nose was no longer sensitive to the odors of snakes, but theirs were.

Finally, I decided I wanted somehow to be more involved in nature and animals, and I moved back to Connecticut. I was about twenty-three, I didn't want to go to college, and I needed to make money. I was by then also involved in body building, jujitsu, and classical music. I felt pulled farther and farther away, not only from my brother, who by then was married, but from the people in my community whom I had grown up with and loved. My interest in reptiles and animals had become more sophisticated and all-consuming.

In Connecticut, I took a job in a necktie factory and stayed almost seven years. I was compelled by a sense of wanting to return to my roots. In a lot of ways I was hurting myself, but I was also helping myself because I needed to have this feeling of belonging. Finally I became the stock manager of all those ties, millions and millions of ties. Then I helped form a union at the company and was eventually elected shop steward. But the first convention was to be held in Miami Beach, Florida, and it angered me that the suite, meals, and car were to be paid for by the money from the union people. I thought I made an honest gesture by using my own money, but when I came back to the factory the workers thought I was crazy. That and other incidents began to make me feel that I had lost touch, and I realized it was time to go. When I quit, my employer told me I would end up a bum.

But what I did was, I went to playgrounds with my animals. I would sit down and talk to kids about animals. After a while the word got around, and people started hiring me. At the time, I had a double yellow-headed Amazon parrot, a couple of crows, a mynah bird, an opossum, and oodles and oodles of snakes, tarantulas, scorpions, and centipedes. But it was a struggle for the animals, with the stress of in the bag, out of the bag, in the box, out of the box. So I went to the Stamford museum and applied for a job. My impression was that I was hired because they wanted a minority person for the minority students. It bothered me to be the token black, but I took the job and fell in love with it because it gave me the opportunity to use all the knowledge I'd acquired over those twenty years. People wanted to know about weather, tides, squirrels — everything. People would bring in baby birds with a twisted beak that the mother bird had thrown out of the nest, and I would rebreak the beak and set it with Scotch tape so it would be realigned. I would feed it with an eyedropper. People would bring in different animals for me to identify, and I enjoyed the challenge of trying to find the answer through books. Children would come, and I'd have them stay, and we would find the answer together. But the museum had cutbacks, cutbacks, cutbacks, and eventually I just couldn't afford to stay there.

I subsidized my income by doing part-time lecturing at Kiwanis clubs, Boy Scouts, birthday parties, and so on, and then some schools found out about me and started asking me to put on assemblies. I would refuse to do assemblies because I felt children should have the opportunity to have a hands-on session, so I would go to a school all day for the same amount of money. This was 1968, 1969 and part of 1970.

Then I heard of a school in Greenwich that was offering a job. I was given incorrect directions and ended up at Greenwich Country Day School instead of Greenwich Academy. They listened to me and liked what I had to offer, and they said, "After you are finished at the academy, please come back and see us." I did and enjoyed myself. Later they sent me a letter asking me to come and teach full time, and I accepted.

My feeling is that when children are born, until maybe ninth grade or so, their minds are like sponges. They want to know everything and have a million and one questions. It seems that after ninth grade the whole idea of school becomes a drudgery to many of them, because a lot of the spontaneous ways they learned in the past, like strolling through the woods, have been taken away from them. My feeling is that you have to be as open-minded as possible in the class and allow every child in that classroom an opportunity to see and hear things the way he or she sees them.

On the first day of school, there are two things I usually say to every class. One is the Eleventh Commandment: "Thou shalt not abuse the earth." The other is that they are not allowed to say "animals," but "the *other* animals." We are animals too. Later I tell them that human beings are perhaps the only animal that can become extinct and the planet Earth would benefit from it. And that to me is just a horrible thought. Yet I believe it is true.

Human beings and other animals must coexist. That is why when I had a lizard in my classroom I didn't cage it. He had free run of the class! Another thing I say to the children is that our greatest curse may be our adaptability. We have gotten to the point where we can tolerate air pollution, water pollution, crowdedness. Perhaps it would be a safer world if we were less adaptable. The kids sometimes think that the animals know I love them, that a dog wouldn't bite me and a cat wouldn't scratch me. And their parents call me Dr. Dolittle.

In my own life, if my wife and I want to go to dinner or a movie and the cages need cleaning, we don't go. I wouldn't want to live next to my feces. There is no reason why they should have to just because I deprive them of the chance to care for themselves. I have altogether thirty animals or so — pythons, boas, frogs, scorpions, tarantulas, a ferret, a dog, and a tortoise. Keeping animals captive is an awesome responsibility, not to be taken lightly.

I see the sacrifices made by many persons far greater than myself in the field of wildlife and animals. I even think I would be willing to sacrifice my life if my death would mean the assurance of the life of all animals. When I go to meetings on conservation I meet people much like myself; we don't really need to be talking to each other. We are all saying the same things —

save the whales, save the birds, and so on. We pat each other on the back and feel good about being together. But it's important to speak out even if other people don't agree. We should take a stand. I am also concerned about the role of black people with wildlife and nature. I went to a reptile symposium in Washington, D.C., and there were maybe twelve hundred people there and only two blacks.

When parents come to school on parents' night with a fur coat or snakeskin belt, or if they say to me that my snake would make a beautiful pair of shoes, I can't let it pass. I am not rude, but I find a way to speak to them about my feelings so they know that everyone doesn't share their feelings. People will say, "What's wrong with animals becoming extinct?" because the animals don't affect their day-to-day existence. I once worked for a school in the Bronx. They just were not interested in snakes. It turns out they were interested in cockroaches, pigeons, and rats. So we began by studying these animals, and they began to understand their own environment. Then I brought in some uric acid droppings from my African rock python. I wrapped them in plastic and used them to write a lesson on the blackboard, just like chalk. In parts of Africa the people do use them as chalk. So the kids became more awed by life. Once when they were fighting in class I brought in one of my 70-pound dumbbells. I said, "If any boy or girl in class can lift this over their head with one arm, you can do whatever you want in this classroom short of killing." They all took the bait and tried. Then I showed them I could and explained that it was because I had a bigger and stronger biceps, deltoid, and triceps than they did. That's how I taught them about muscles.

One summer I was working at a day camp for children, teaching about nature and animals. I became like a Pied Piper; they were in my lap, all over me. It was a lovely feeling, but I also realized the terrific responsibility and impact one could have. On the last day of camp, this little girl about seven years old said, "Don't go home till I tell you a secret." Finally the buses were coming and she said, "I can't tell you here." I said, "Where?" She took me behind a school building and had me bend down, and said, "Jim, I love you as much as I love God." I said, "What?" And she said, "God must be your best friend because you are good to all the animals." And she gave me a big kiss on my cheek and I was left — speechless. It just blew me away!

Parrot Patron

Hugh and Arturo Vandervoort live in the suburbs of Annapolis, Maryland. Hugh is thirty-eight, divorced, a special agent at the Prudential Insurance Company of America, and a sometime foster parent. Arturo is less than half Hugh's age and is a parrot. Hugh and Arturo are almost inseparable and have become minor celebrities in their home town. Part of this fame comes from Hugh's excellent sense of humor. He says that Arturo has one too, but is just less talkative.

Hugh and Arturo live in the upstairs of a split-level house; a third roommate lives downstairs. Arturo has the run and flight of the house. His perch is a custom-made pedestal with a tree branch and large chains for climbing. He also has a big birdcage in the bedroom, but frequently sleeps in bed with Hugh. In the yard are several tall trees, which Arturo likes to fly to or climb up. This necessitates the presence of a large ladder, which Hugh often has the opportunity to use with Arturo's encouragement.

HUGH VANDERVOORT: I don't think anthropomorphism is a viable concept in this house. Arturo *is* a person. My house has suffered from fifteen years of parrot infestation. He's torn up the drapes and left little footprints here and there. The house rule is: If something can be turned over and he can get to it, it will be turned over.

The truth of where he came from is, they get kids to climb the trees in the jungle and check the parrot eggs. When the babies have enough feathers to be protected from the elements but not fly, they'll take them and put them in cages. The kids chew up corn, vegetables, and fruit and feed the baby parrots from their mouths. This is the way the mother feeds them. Arturo still likes to eat that way.

But one day on May fourteenth, on his tenth birthday, he and I met two girls in a bar. I had a few more than I needed, and I made up another story of how I got him. I said I was in the Peace Corps down in Guatemala when this palm tree fell over and a little egg rolled right at my feet. Just then it hatched and of course he imprinted on me, and I've been having a birthday party for him ever since. That last part is true, and each year it's a different party. This year was tuxedo and sneakers. One year I had invitations engraved. Another, I called people. It depends on my mood and money.

The guy I bought him from was a bird lover. He would hand-tame every one. His hands were always bloody, but he refused to use gloves or sticks or anything because he was convinced (as I am) that it's the only way to tame them. This guy wanted to get quality birds to the public at a reasonable price.

He loved all his birds. There was no crowding, and he had a cockatoo for forty years that had a thousand-word vocabulary. I felt rapport with Arturo right off, but I visited him three or four times before I bought him. When I bought him, the man's wife gave me a beautiful cloth for a cage cover. She told me Arturo was the nicest parrot they'd had in a long time.

Arturo was the only name he knew other than Roberto, and I liked Arturo better. I tried for hours and hours to teach him anything, even curse words. I believe he can do it but just won't. He's a good singer and likes Jimmy Buffet, but mostly he just hums along with some rhythm. He'll hit one note and go off on his own. Arturo told me that ever since Latin became a dead language, he's lost interest in speaking.

I think people can get a lot more companionship from a parrot because of the longevity. The relationship can grow. Ours has grown over the past fifteen years. Arturo's a yellow-naped Amazon. I took him to a biology class for little kids. I wrote his name, *Amazona auropalliatus,* and it took up the whole board. I've done a lot of reading about parrots and have learned a lot from him. When I first got him, I treated him like a China egg; then I realized he was pretty tough.

I was born in Chestertown, Maryland, and moved to Ohio, Texas, Florida, Mississippi, New York, and then back to Maryland. I always liked animals. We had a dog that lived till it was fourteen and got run over by a car. I got the remains and dissected it. My mother was horrified. It was a blond cocker spaniel named Tinker. My father was a college professor, and Tinker followed him to class and got in the paper a couple of times for wearing a mortarboard and a pair of glasses for graduation. My father was a quiet guy, but you could tell Tinker had a special place in his heart. There were five of us kids.

When I was thirteen we were in Mississippi, and my dad was diagnosed as manic-depressive. He ended up dying of a brain tumor, and the surgeon said that could have been his problem all along. My father had sold insurance as well as being a professor. I always swore I'd never go into that business. I used to work as a secretary in the Department of Agriculture in D.C. One day I just quit. I couldn't stand it. I took a boat trip for six weeks to get my act together and got into selling insurance. I've been doing it ever since. My college degree was in business and economics, so it's kind of natural. People appeal to me, and I'm not shy either, which really helps. I'm the only one in my office who hired a secretary. She knows Arturo, but does not like the fact that he eats his pencils on *her* desk.

I got tired of people asking me "Does he talk?" and "What's his name?" So I made him a card saying ARTURO VANDERVOORT, M.T.A.J.D., which stands for "Meaner Than a Junkyard Dog." I had to have his phone number changed because he was getting too many calls. When I go places with him, 145 people actually come up behind us and pull his tail. I pinch people on the ass when they do that and say, "Do you like it? Well, he doesn't either." I was surprised at how angry I get when someone does that. I'd much more likely retaliate in a physical way if someone messed with my parrot than if they messed with me. The attachment is probably even deeper than I think. Sometimes I think that a lot of people are not worth half as much as a good parrot.

There are a couple of local bars that know me, and I go and sit in a corner with him. People get all excited, so one day I just extended my trip to every bar in town. And it's fun riding down the highway with Arturo on my shoulder. People do a classic double take. I have even gone roller skating with him. They assigned two attendants at the rink to skate around with me to make sure he didn't bite.

People's attitudes are funny. Most don't have the faintest idea of how to deal with a bird. They come up and stick their finger in his beak, and don't realize how he is perceiving this giant thing with a big rumbling voice that is coming after him. So they get bitten. I've got a couple of friends who are absolutely terrified of him. He was on the top of the steps one day, and this guy opened the door without knocking and surprised him. So he flew at his face. After that, my youngest sister, who finds me parrot things like the parrot potholder on the refrigerator, got a sign saying WARNING, THESE PREMISES ARE PATROLLED BY AN ATTACK PARROT.

Arturo likes to go places with me. Actually, it's hard to get somebody to parrotsit for any length of time. One time he was with me and got his own library card. I went to the library with him and the librarian asked if he had a card. I said, "No, but he needs one," and she told me to write to the supervisor and explain his problem. I had some stationery with his name on it and sent an autographed picture. It was pretty straightforward. The library administrator replied promptly. He regretted that Arturo had been refused a card, calling it "a bureaucratic foul-up," hoped that future service would be satisfactory, and thanked both of us for our interest in the library. They did a newspaper article on it afterward.

He likes to go to the office with me, and he chews his pencils there. There's a rental car company that's always trying to solicit our business. I told them I would recommend them if they would keep Arturo supplied with pencils. So the salesman sends me forty or fifty pencils every month, and Arturo destroys them. He treats them just like a bone. The lead is the marrow. He splits them lengthwise and pulls it out.

As a joke, I have a million-dollar whole life policy listed in our computer on my "son" Arturo. I had to record him as age sixteen, and the premium reads as $12,000 per year! That's for the first forty-nine years. After that he can retire at age sixty-five at a monthly life income of $20,906. Two years ago

I sold $3 million worth of life insurance. On a past computer information form, I had listed my wife's name as M-U-D and my son's name as Arturo. When the news clip came out it said Mr. Vandervoort lives in Arnold, Maryland, with his son, Arturo, and his wife, Mud. I had them leave out the wife's name.

Arturo has helped me in my business. One day I was at a guy's house, trying to sell him insurance. He had a parrot sitting on top of a cage in the kitchen. The guy said, "I'd put him in the cage, but every time I try to put him in he bites the door or hangs on to the side." So I picked up the parrot and turned him backward and put him in tail first. The guy bought the policy. He said, "I don't care if it's a good policy. You just saved me so much trouble, it's worth it."

There was a fund-raiser at Children's Hospital, and the director is named Dr. Parrot. So Arturo sent him a check (he's on my personal checking account) and a note that read: "It is great to see one of our kind getting ahead in the world." They published it in their paper, saying, "We even take checks from parrots." Arturo even has his own American Express card.

He's gotten very territorial about certain places in the house. A guest shouldn't try to go in the kitchen. They will be safe from his attention as long as they don't move. If they do, he attacks. When he dramatically changes the size of his pupils to focus, it means he's mad. He looks ferocious and does a little sailor dance, spreads his wings slightly, and holds his tail out. Then I will not pick him up; if I do, he'll bite real hard. He's ready for business. He watches birds come to the feeder. If a bluejay comes, he'll get aggressive. If a chickadee comes, he won't even notice. He also gives a warning. It took me a long time to figure it out. He will walk up to you, stop short of your feet, and bury his beak in the rug. He'll do it a couple of times. But people don't understand that it means, "Why don't you back the hell off because this is my rug." I've even seen people stick a toe out to him with disastrous results.

I had a roommate a few years ago. One morning I was out drinking coffee and he was asleep in his room. Then there was this scream. Evidently he had left his door open just a crack, and Arturo can open doors. He pushes them by putting his beak on them and leaning forward with his feet. He climbed up on the guy's bed and probably watched him sleeping. As soon as the guy moved he got him.

Early in our relationship I learned to scratch his back and work my hand down to his wings. At the wings it was hands off. At about that time he started wrestling with me. I thought he was playing, but decided, "That's nonsense. Birds don't play for the sheer joy of it." But he'll practically roll over in my hand and play with me with his beak and feet. With Arturo I can have a relationship without reservation as opposed to one with people. I'm a pretty friendly person, but Arturo brings something to a relationship that people can't. Like when he rolls over, it's not a sign of submission, it's an absolute trust: "Here's my bare belly. Do whatever you want with it."

Parrots are very destructive. I have to watch Arturo all the time. In the foyer there's a ledge. If you put things out there in a row, he'll push them over and watch them fall and bounce, then go on to the next one. If you keep putting the things back up, he'll get real mad and push everything off without bothering to look anymore. It's funny to see.

I've driven a lot of people away when Arturo and I are having dinner with them. If you don't give him food from your mouth, he'll go after it. He'll probe with his tongue in your teeth. It's a funny feeling. It's kind of rubbery and strong.

He'll eat just about anything I will. I'm very surprised at the gamut of things. He'll drink wine, but not any straight whiskey. He shouldn't have too much fat, but if you give him a piece of meat with fat on it, he'll go right for the fat. Or if anything has melted butter, he'll eat the butter first. He eats marijuana seeds too; somebody gave me a whole bunch, and I was surprised he could eat such tiny seeds. Someone told me that some imported bird seed has sterilized marijuana seeds. Of course, he eats peanuts and sunflower seeds. Sometimes I grow him sunflowers. I think one reason he's done so well is, he gets a little bit of everything. Someone gave him a can of bird seeds for his birthday. It says, BIRTHDAYS ARE FOR THE BIRDS.

Then there are upkeep and maintenance rituals. If you want to snip his nails it takes two people, and the other person has to be unafraid. If he happened to bite during that, I'm sure he'd break a bone. He hates to bathe. A real ritual is beak cleaning. You learn to keep things around for him to clean his beak on, otherwise he's going to clean his beak on anything. You can't have good furniture with a parrot. But you learn to leave newspaper, sticks, and pine bark for beak cleaning. And he doesn't do a cursory cleaning. He takes a little piece of paper and runs it outside the top of his beak. Then he clears the ridges inside his upper beak, just like cleaning between your teeth.

Parrots molt once a year. I know it takes several days, but it always seems to me that all of a sudden he looks ratty as hell. There are big gaps in his tail feathers. The wings don't look so obviously bad because they are folded, but when I stretch them out they look just awful. The molt is very disciplined. First, his big flight feathers go. Then the outer two tail feathers, and then the two inner tail feathers. That's so they don't lose their ability to fly. There's a lot of down that comes with it. For a couple of months there is down everywhere. I go to work with down feathers in my hair. I don't know if he molts every feather on his body, but sometimes when you look in his cage it looks like he has.

I compare molting to a menstrual period, but only once a year. He gets a little irritable, and he slows down. When his little quills start to break through, they must itch like crazy. You have to be careful when you are petting him because the quills are rigid and if you brush the wrong way it hurts him. He will sit in the corner for hours, looking like he doesn't have a whole lot of energy to do anything else. I try to make sure he gets more protein, more vitamins, and more green vegetables. Because I like him so much and because I'm with him so much, it seems to take forever, but I would say he goes from start to finish in six weeks. Then he looks terrific. It's really unbelievable. The old feathers were worn and abraded, and suddenly he's all new.

I look back at the changes in my life. I used to shoot sparrows when I was a little kid. Dumb stuff that I would never dream of now. I think going to Vietnam might have had something to do with it. Death is real. It doesn't deserve any modifiers; there aren't any that apply. I was there in '67 and '68. All I did there the whole time was fight. I was an infantry platoon leader. It was part of growing up, accelerated. When I went I felt, "Okay, I'll defend my country like my father did." But the more it came out that it was an economic and political war, anything except a ground-gaining war, I began to feel there was no point to it. The only purpose was to stay alive.

I was there eleven months and fourteen days. As soon as I got back to the States my wife got a parakeet. I was going to school on the GI Bill. And I felt as if I was older than anybody around me. But I was more impulsive than before I was in Vietnam. I mean, the $250 for Arturo fifteen years ago was a lot of money that we couldn't honestly afford. My wife was damn angry. But especially the first few years after I got back, I lived to grab all the gusto. A bird like Arturo nowadays is $750 to $1000.

My wife never really got along well with him. He would chase her around the room. I think she made the mistake of letting him do it. After I had been married for ten years, my wife said it's me or the parrot. So he was on the invitations to my divorce party: "Arturo Vandervoort requests the honor of your presence at the divorce of his parents, Jane and Hugh." Jane wouldn't come. That's why I got rid of her — no sense of humor. The only thing that makes me more angry than someone messing with my bird, Arturo, is when I hear people saying anything good about Richard Nixon and Henry Kissinger. Anyway, I was never really good at the old school idea that said that a man doesn't show emotions and never cries. It's easy to show my feelings with Arturo; there's no criticism.

Photographing the Untamed Image
of a Living Soul

I was told about Lois Constantine by a woman who was living alone in a garden apartment with ocelots and margays. Her husband had told her to choose between him or the cats and she had. She told me that Lois was a person who knew "all the animal people." So I called her, and I found that she was the hub of the wheel of that large subculture in the Los Angeles area.

Lois is an animal photographer. She is divorced, childless, and in her forties. She lives in a residential neighborhood in the San Fernando Valley and shares her home with a Great Dane, a Doberman pinscher, three cats, and two Indian ring-necked parrots.

Her photographic studio and darkroom are in the house, and a variety of clients including Hollywood stars have sought her skill to capture the essence of their pets.

LOIS CONSTANTINE: I was born in Chicago, and at that time my family was very poor. My father was just beginning the Comet Model Airplane business, which became one of the largest model companies in the world. By his early twenties, he had earned enough money to buy a custom-built Georgian colonial home, with servants to take care of my younger sister and me. But I seldom saw my parents then; I have few recollections of family activities. I was always a quiet and shy child. I remember, when I couldn't sleep I would go into my closet. On the door was a mirror. I would leave the door open just a fraction, and there would be patterns on the mirror. I would imagine them as lions, tigers, and all the jungle animals, and they became my friends. So as far back as I can remember, wild animals were something beautiful and awesome, but not to be feared.

I had always wanted a pet, but my parents wouldn't agree to it. Finally, when I was eight years old, we moved to a suburb of Chicago where there was a large yard. At that time I was given a rabbit. I was thrilled. I played with it every available moment. But one day I came home from school and raced into the yard, and to my horror she was gone. My parents had disposed of my prized pet with the excuse that they feared I would hug it to death. Later on, I realized that that day began my bonding and loyalty to animals. I don't remember ever thinking that animals were less than humans and I was always curious about how they felt, what they thought, and how the world seemed to them. Animals made me feel less alone. Somehow, in relating to them I learned how to relate to people.

I married a man I met during a vacation in Greece. We bought a home and got an Australian shepherd, and later a Great Dane. Over the years I acquired several cats that I rescued, and we built an aviary with canaries, finches, doves, and a section for parakeets. I also did rehabilitation and foster home work for many small animals, including ocelots, baby tigers and cougars, birds, and tortoises.

I have been an artist for most of my life and earned money exhibiting and selling sculpture, weavings, watercolors, and crafts. About thirteen years ago I ordered a tapestry loom from Finland, but there was an eight-month dock strike. Just to pass the time, I enrolled in a photography class at a nearby high school. I instantly fell in love with it. I arrived early and left late and used most of the paper and chemicals. I was hooked. We learned to make salon prints for exhibition, and I was encouraged to enter shows. Having never been competitive, it amazed me that I continually won. Those successes motivated me to think of photography as a profession.

From the beginning, my passion was to photograph exotic animals. My first were a cougar and a leopard, owned by the same person. The owner kept shouting warnings because he was afraid the cougar was going to bite me. I didn't feel any fear, just an awe that this wonderful wild creature would share itself with me. As I was photographing it, I was longing to touch it. After I finished shooting, suddenly to my delight it just jumped onto my lap. Then I photographed the leopard. He was caged, and the shot I got was a portrait of a proud animal behind bars with one tear coming from his eye. I wanted to capture its sadness. It was the trapped, locked-in, isolated feeling that I understood.

Early on, I made a decision to capture the gentleness of an animal, never its violence or aggression. It takes time to find the right moment, but if you wait long enough they connect with you. It's that connection I look for — an eye contact, even body contact. We human beings assume we can take whatever we want from this planet. Wild animals retain their independence and only at times allow you the privilege of entering their world. Since I have always been inclined to capture something of their personality, my work tends to be more portraiture than full body, though I am fascinated by the exquisite graphic art of their patterns.

There is a memory of an animal that haunts me. A zoo animal, a jaguar that has since gone. She became old and wasn't a perfect enough specimen, so she was sold to a zoo in Mexico. She was one of my favorite cats. There was a sadness and resignation about her. I have a picture of her looking over her shoulder with the most forlorn look. It seemed to reflect the tragedy that life can be. I'd like to think of that animal being free to roam her territory, not imprisoned in a small enclosure. She was magnificent in her beauty and power. Over the years she had babies that I would photograph with her.

When she was gone I cried. They never honored the dignity of her growing old.

Photographing exotic animals is not without risk. I have had one bad experience, when my index finger was almost severed by a big cat. It was done in a playful way, without intent to hurt me. My greatest hurt was that I lost some of the trust I had had before. Also I was fearful that I wouldn't be able to use the finger to take photographs, but I decided that if it didn't work, I would build an apparatus or use another finger. There was no way I was going to stop taking pictures. It was almost a year after the accident before I could begin to trust big cats enough to photograph them. If a person had injured me, maybe I would have felt bitter, but because it was an animal, I was determined to overcome my grief.

I view everything as though I am looking through a lens. Many times that image gives me a sense of great joy. Photography is my way of communicating what I feel the animals communicate to me about themselves. There is always that moment of truth with each and every animal, a kind of vulnerability that reveals the true spirit of the soul. On a deeper level, I suppose I can identify with the struggle of the captive animal fighting to maintain its identity as a wild and free creature. It gives me chills to think about it, because what I photograph is something no one can ever take away — the untamed image of a living soul.

Johnny Carson's Animal Person Loves Horses

Joan Embery, thirty-three, is a goodwill ambassador for the San Diego Zoo and has appeared many times on Johnny Carson's *Tonight Show* as well as on other programs.

Her personal animal preferences are horses and elephants. The 20-acre ranch she shares with her husband, Duane Pillsbury, and his teenage daughter, Holly, has barns and exercise areas for her twenty-five horses of many different breeds.

JOAN EMBERY: Growing up in San Diego, I thought I wanted to be a veterinarian. We had springer spaniels. Mom worked for a veterinarian for a number of years, and my uncle is a veterinarian. People always ask me when I "discovered" animals, but it was always there. I always had very strong feelings toward animals.

I started riding in junior high school. I begged my parents and said I would do anything for lessons. I took them regularly, then got involved in showing horses. I started with saddle horses and gaited horses, and then 153

went into jumping. I got interested in dressage and cross country, then driving horses.

Probably one of the reasons my interest in horses is so strong is that I enjoy working with animals, and I enjoy seeing a result from all the hours spent with them. With a horse, you can train it to jump over a 4- or 5-foot fence, or train a team of horses to work together. There's a rapport and understanding that's a give-and-take and is a physical, intellectual, and emotional challenge. The thing I like about working with horses is taking your own intelligence and abilities and combining them with the abilities of an animal and making teamwork.

At eighteen, I started working at the San Diego Zoo as a Children's Zoo attendant. I worked with young wild animals — cleaned up, fed them, and answered people's questions. I had spent a lot of time at the zoo when I was younger, but I never thought of making a career there. I liked the elephants best, and worked toward ultimately doing elephant shows myself.

I started working at the Children's Zoo at the same time that Carol, a baby elephant, arrived. She was my favorite animal, so I spent extra time with her. She and I did shows in the zoo, and I trained some other elephants and did shows with groups of them. My first appearance on national television was on the *Tonight Show,* which came about through Carol, after I had trained her to paint with a brush. I think that the appearances on the *Tonight Show* have been successful because of the spontaneity and naturalness. A prime example would be the night they wanted a koala on the show, because everybody in the world is in love with koalas. We took a koala and also a common marmoset. The koala was the big draw, but the marmoset stole the show. Johnny Carson and I were talking about how marmosets come from South America and that they are arboreal and feed high up in the trees. At that point it jumped on top of Carson's head, and he said, "Well, you can't get any higher than that." Then it urinated! I said, "Well, you know when they are in the trees they mark their territory. And that's what he's done." Everybody was amused, yet it really said something about what marmosets do in nature. It was played again and again on anniversary shows. When I go to any city in the United States, it's the first thing people bring up. And nobody has ever mentioned the koala.

Initially, I was an animal collector. I had ducks, chickens, and wild birds. But as time has gone on, I don't have to own an animal to feel close. I think you have to appreciate animals as they are. Now we have fifty horses on the ranch and own about twenty-five. There is a team of Percherons, a Clydesdale, miniatures, a thoroughbred, quarter horse stallion, Shetland ponies, some Arabs, and a team of circus Liberty horses. Most people tend to specialize, but I'm interested in so many I can't seem to pick just one.

I like draft horses because they are kind of a lost art. For years they were   155 all but forgotten, but they are really magnificent, with their size, power, and temperament. I like the miniatures because they're cute. You can take them everywhere and they become pets. I like the thoroughbred because he's highstrung and very athletic and strong. The quarter horses are just nice, good, pleasure and work horses. The Arabians and Liberty horses I use for tricks.

I've heard it said, and I tend to agree, that every time you work with a horse you are either training or untraining, because what you do and how you do it and how you handle yourself are either conditioning the animal to respond positively or not. And they are all different. My black stallion quarter horse is the kind of horse who will try to work for you. Melody, the Clydesdale, is lazy. You have to push her a little. But if I needed a horse to pull a vehicle through downtown traffic, I would choose her because she can handle stressful situations well. The horse I am closest to is my Arab. He's special since he's my first. I call him Finally, though his registered name is Rajrahmoun. I raised him from four months and stayed up nights and slept with him when he was sick as a weanling foal. He's the kind of horse that when I come home at night he'll turn around and whinny at me. He knows me well and there's a closeness there.

It's frustrating sometimes because my true love is being with the horses, and I find myself being put in a situation where I have less and less time. I get tired of being interviewed, tired of all the demanding television work, and tired of being put in a situation to do things that don't show the animals at their best. You can get a producer who will expect the animal to come out and do a back flip or expect the animal to fit into some unnatural form. It's frustrating sometimes.

I feel you have to reach people on their level. You can do an educational show and reach hundreds of thousands of people, or you can do a primetime network talk show and reach millions. That's what it's all about — reaching the greatest number of people and being able to create a fascination for animals that may not have been there. It's amazing the number of people who have come to the zoo as a result of the exposure on the *Tonight Show*.

When Duane and I met we had no intentions of marrying. Some people said, "He is twenty years older than you are," or "She must have married him for his money." That was frustrating because I thought of myself as being totally self-sufficient and independent. Duane and I look at people with their Mercedes and their beach houses, and we kind of smile at each other and say that with all the money we have spent we could have anything we want, could retire, or live anywhere. But we wouldn't be happy without the horses, the other animals, and the ranch.

Holistic Veterinarian

Sheldon Altman, D.V.M., is one of the pioneers in the use of veterinary acupuncture. The classical oriental theory of acupuncture suggests that the life energy known as Chi flows as positive and negative components through channels in the body called meridians. Where the channels come close to the surface of the body, they create acupuncture points. Each meridian corresponds to a particular organ. Disease is seen as a process of disruption of the flow of energy, either too much or too little. Acupuncture, by stimulating particular points, adjusts the energy levels back toward normal. Acupuncturists have also explained their technique of pain relief with models from Western medicine. The theories use neurological ideas of the blockage of pain impulses in the spinal cord or brain stem, or by stimulation of the autonomic nervous system.

Dr. Altman works at the Animal Hospital in Burbank. The hospital's mascot is Chu Chu, a seven-year-old black and white Shi Tsu, who was brought in two years ago with complete paralysis of her hindquarters secondary to spinal cord damage. She had no control of her bladder or bowels. Rather than choose euthanasia, Dr. Altman, who felt "she was too vital to be put to sleep" purchased a specialized chariot-like device which has two loops for the hind legs, rods that go over the back, and wheels that extend to the side of the hind legs. Chu Chu wheels herself around the clinic, and a staff member twice daily manually expresses her bladder and cares for her bowels. At night Chu Chu goes to the edge of her basket and pulls herself out of her contraption into bed. Dr. Altman says, "She owns the hospital and we work for her."

SHELDON ALTMAN: I have been a veterinarian for a little over twenty-one years. I graduated in June of '61 from Colorado State University. I think I wanted to be a veterinarian from the age of twelve. I grew up in Colorado in an environment full of animals. My father was a farmer. We had 160 acres, and about 40 acres of that were in corrals where we fed cattle for beef. There were dogs in the house and cats all around as outdoor animals.

Even though I was raised with a farm background, I admit it did bother me on Sundays when we decided which animal was fat enough to go to market (Mondays were the market days at the Denver Stockyard, when calves I had become attached to went to market.) I couldn't eat beef for weeks because I would be afraid that I would end up eating my friends. I really felt self-conscious about it because you are not supposed to think of these animals as individuals. Still, I think that on farms, everybody has a relationship 157

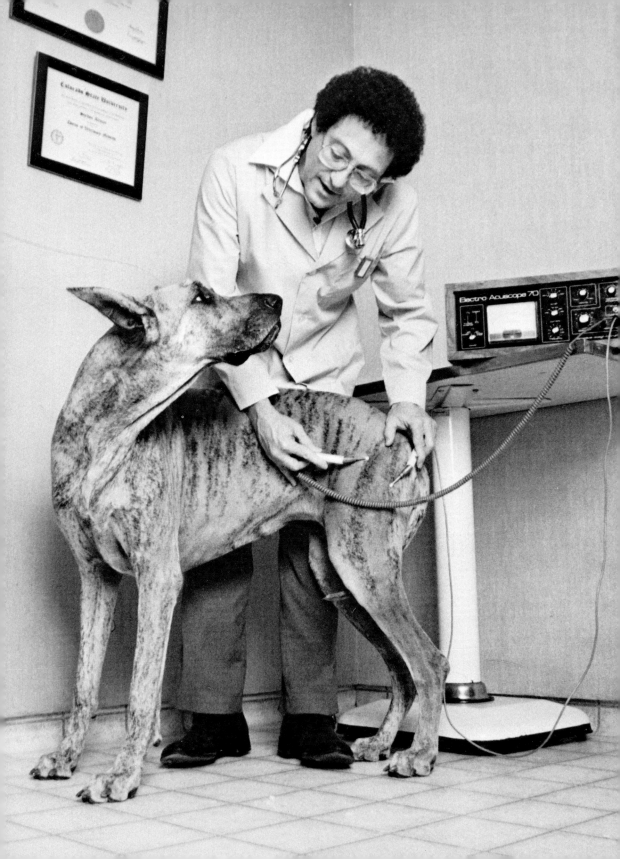

with the animals. I've seen people out there in storms trying to warm up calves, or doctoring them when they're sick, or feeling for them when they are in pain. Yet ultimately they know what these animals are being raised for — somebody's dinner table.

My dad is a remarkable person. He could sit on horseback and look out over a corral full of cattle and tell what kind of physical condition they were in. He could pick up disease more rapidly than I could, even after I got my veterinary degree. Actually, he did a lot of treating by himself before we called the veterinarian. He knew which antibiotics were effective for certain diseases, and he could give intravenous fluids, but he knew his limitations and treated with the advice of our local veterinarian.

I think one of the reasons I got interested in being a veterinarian was the fact that I went through a very heavy trauma with one of my dogs when I was twelve. He was a little black and white mixed terrier named Tiny. At that time we didn't have as effective vaccines as we do now. The vaccines were killed virus products, and sometimes they gave immunity and sometimes they didn't. Tiny had been vaccinated, but he came down with distemper. He went through a long-drawn-out process of making clinical recoveries and coming down sick again. I did a lot of TLC with the poor little guy. I had an aunt and uncle who lived next door to us. They had no kids and were very involved with this poor little puppy too. He started out with the typical gastrointestinal problems, vomiting, and diarrhea, and went to respiratory problems, with pneumonia. Finally the virus invaded his central nervous system and caused encephalitis. Eventually, he had to be put to sleep. He had slept on the foot of my bed, and even after he was gone I could almost feel him on the bed for three or four months. It really left a lot of frustration, and I wanted to get into a field where I could alleviate things like that.

The veterinarian who took care of Tiny made an impression on me. He showed empathy for me and the dog. I think that kind of grabbed me. He was in his late twenties or early thirties. He had a brusque manner, yet wasn't abrasive. He really was able to feel sympathy. Later, I worked for him scrubbing kennels. Our farm veterinarian had taken me on calls with him, and I worked for other veterinarians with livestock, but I stayed with that veterinarian who cared for Tiny on and off until I actually started veterinary school.

When I started school, I wanted to be a large-animal practitioner involved in beef and dairy cattle and food animal medicine. But I realized that it would not be economically feasible to do heart surgery on cattle, for instance. I felt in small-animal medicine I would have a freer hand to run laboratory tests and do more complicated diagnostic and surgical techniques. One of the things in large-animal practice that bothered me, especially as far as food animal practice, is that you are dealing with a product, so to speak.

It becomes a matter of economics, and the animal is not thought of as an individual.

After I worked in the field for a while, I found there were people who had gone to veterinary school simply because it was a way to make a living. They didn't get emotionally involved and were pretty effective as far as reading test results and picking the proper treatment. But I felt there was something lacking. Sometimes the difference in whether an animal makes it or not is the will to live. I really think an animal can sense when you are with them or whether they are doing it on their own. I can't fault the technicians, but I think the finest veterinarians are people with the extra element of personal involvement.

My career took a turn when I got into acupuncture in about 1975. I began as a skeptic. I saw it as a kind of quackery at best, and at worst as a ripoff of the public. But I was stimulated to at least think about it by a lecture in San Francisco at the American Animal Hospital Association Convention. It was given by an M.D. from Salt Lake City who had gone to China and had seen how acupuncture works on people. I began to think about using it on animals. There were some people at UCLA at the time who were beginning an acupuncture project with animals. They decided that if it could be shown that it worked on animals, it might destroy the skepticism. They set up five clinics in the Los Angeles area and encouraged local veterinarians to send their refractive cases. I didn't send cases at first because I thought it was a strange method of treatment. Then some of my clients demanded to be sent, and I realized that cases that weren't responding to regular medicine or that were inoperable came back quite improved or even cured.

Several veterinarians, intrigued by the clinic's results, formed an association called the National Association for Veterinary Acupuncture and prevailed on two of the acupuncturists from the project to teach veterinarians what they had learned during that year. A colleague and friend asked me to take the course with him. After a few sessions he decided to drop out, but I was hooked. I took another series of courses and found it a fascinating new way of looking at things. I also took courses at UCLA and worked in the UCLA pain control unit, and I accumulated more than two hundred hours in two years watching their human acupuncturist and assisting in treatments. I didn't actually insert needles in people, being licensed to work only on animals, but I could stimulate them after they were in, log the points in the medical records, and remove the needles.

When you are a student of acupuncture, you start by finding points on yourself, and you realize by actual sensation that there is a difference between an acupuncture point and a nonacupuncture point. At a point you get a little buzz, almost electrical, sometimes almost a shock. After the needle is in, there is a kind of heaviness at the point, almost as if you have a BB in your skin. It is a dull but not unpleasant ache. If you stick a needle in the wrong place, it can be painful. There are traditional points that have been handed down for thousands of years. There are new points that have been discovered electronically. There are points found by transposing human anatomy onto animals. It takes a lot of experience and practice, but after a while you instinctively locate the points. In most cases they are in depressions, and you can develop a touch where you can feel the points.

I used acupuncture sparingly at first. As I got more comfortable, I began using it more and more. Now I do between thirteen and sixteen treatments a day, and it's a big part of my practice. I use it mostly for chronic and degenerative conditions. I see a lot of hip dysplasia, arthritic backs, disk disease, and chronic skin conditions. The acute conditions usually respond nicely to Western medicine, and there is just not the need for acupuncture. But this is exactly opposite to how acupuncture is used in the Orient, where it is used as a first-line treatment, followed by Western medicine if that doesn't work.

Two years ago we pulled two hundred and sixteen cases from our files and found a 63½ percent success rate. I wasn't too favorably impressed, and in fact I was a little depressed. I thought that if I did surgery at that rate I would throw my scalpel away. But then my wife said, "Why don't you look at what you are treating?" Sure enough, the cases I had been treating had been refractive to almost everything and were on their way to euthanasia when they were referred to me. So that percentage on a salvage rate was really not too bad. They could live a comfortable if not entirely normal life and would still be satisfactory as pets.

Acupuncture leaves me more in tune with the fact that I am treating a part connected to a whole animal. It also makes me consider the ecological situation of how the animal relates to its owner, its family, its environment. I have a much more holistic approach to my medicine.

The treatments can change the behavior of animals. A client will come in and say he hadn't realized it, but his animal was acting like he did two years before his problem developed — he was relaxed and playful. Many times I can see an attitude change before I see a change physiologically. You can see a change in the expression on the animal's face. His eyes look different. But I'm a masochist. The cases that stick with me are my failures, the ones I see day after day and when I go home I'm still trying to figure out why they aren't working. But I do get some of the miraculous cures — one or two treatments, and they are fantastically better.

I don't know if my patients appreciate my efforts, but I do know they respond to a friendly approach. If you go in there aggressively, you can destroy the relationship with them. I give them treats and petting and give them verbal rewards so they are not frightened. About 60 percent of my acupuncture cases are referrals from other vets. When they first come in, they and their owners are tense and on guard. The owners don't want to build up false hopes and are worried about financial output, and the animal is like a big emotional mirror and comes in like a bundle of nerves. So the first visit is based on explaining what we hope to achieve and trying to make friends with the animal. It is important to do this because the more nervous the animal, the less effective the therapy.

The area of veterinary medicine I have the toughest time with is euthanasia. I think that probably I have kept some suffering animals going a little longer than I should have. In veterinary school you get the idea that if you make the correct diagnosis and use all the correct procedures and medications, you can't help but effect a cure. But when you get out in the cold cruel world, you find you can do everything by the book and still fail. But I find I can't buy it as anything but a personal failure on my part. So a lot of times I think, "Maybe tomorrow I will come up with something better. Let's keep him alive another day." But finally I have to realize that I am looking at a really terminal situation, and the animal is suffering and we better get him out of his pain. I have to do it, but I still get a cold chill in my stomach when I push the plunger on the syringe. Sometimes I have the owner with me or a technician so somebody is with the animal. I feel as though I am going down an elevator shaft inside. It's an odd and terrible feeling to look a creature in the eyes and it's alive. And as you are watching it, in an instant it dies. It's like blowing out a flame. It's a scary responsibility. The decision is based on what I think medically and what the owner feels emotionally. I am not a hired killer, and I won't euthanatize an animal out of convenience. If somebody is leaving town and they decide, "Hell, let's just kill the dog," I will not do it.

Animal rights have not been legislated yet, but the concept is creating a lot of interest and discussion. I do feel that any ethical or moral background demands that you respect the fact that this is a living individual, that it can feel pain, and that it cannot control its own life. I am a religious person, and I believe the Bible gave us dominion over animals. Dominion implies responsibility, not just dominance. We have the obligation to see to their well-being. I see it this way: If I am going to make life-and-death decisions, and the animal has no input of his own, I had better consider what his input would probably be if he had an intelligent choice.

The Lady and Her Tigress

B.C. (Beautiful Cat) and Dee Arlen live together in a mobile home on an acre near Riverside, California. Dee, who has worked as a commercial designer, has furnished the interior with a crystal chandelier, velvet-upholstered walls, and candy-striped velvet furniture. The bedroom has gold velvet walls and a bedspread of quilted tiger print.

B.C. uses the rear door to enter her large yard, which is enclosed with heavy chain link fencing. B.C.'s yard looks like a jungle, with trees, shrubs, and flowers that Dee planted for her. The litter box is an old sink, and she has a heated den box for cold days and a large fan for hot ones. She also has an automatic drinking fountain and her own radio.

DEE ARLEN: I learned young that we are not the victims of fate or destiny but of our appetites. It would seem that mine have turned out to be a tigress and horses.

I grew up in the Midwest, in a spacious house (not to be confused with a home) with servants, a nanny, and luxuries only the wealthy could afford in those post-Depression days. During my childhood I had everything money could buy, but not the most important things — love, understanding, and a meaningful relationship with my mother and father. Just my nanny gave me love. My parents divorced when I was seven. Deep scars were left by this transition because I was a pawn in their chess game called divorce. One side of the chess board was my father — and my horse. On the other was my beautiful, aristocratic mother. During the divorce my father said, "You can't live with your mother and have your horse, so you decide." What a decision for a small child, especially when my nanny saw to it I could ride my horse all the time, and I forevermore clung to my horse as a security blanket. When the judge asked who I wanted to live with, without hesitating I exclaimed, "My horse." So I ended up living with my father.

My father's fuel was wine, women, and song — in fact, he was a proverbial playboy — but he gave me the luxuries of charge accounts, cars, travel, and horses as a teenager. I also grew up with Great Danes and always had a harlequin. My dad had a black Dane who slept in a twin bed with him. That's no worse than my sleeping with a tiger now, I guess. There's no more room in a twin bed for a dog and a man than there is for a tiger and a lady.

He gave me expensive American saddlebred horses and lessons from a 163 distinguished trainer. Only the horses he bought for me were truly mine. He didn't know or want to know anything about horses, yet he owned race-horses; it seemed to me he just paid for them. But because of the racehorses, I learned everything about caring for a horse. However, I am grateful to my father for three things. He taught me: to dare to be what I am, to resign myself with good grace to all that I am not, and to show unswerving fidelity to whatever I believe in. Because of Daddy, I became a rebel, a dyed-in-the-wool maverick, and thus I have remained.

I attended a finishing school — "just to soften the tomboy," Father said. After I went off to Ohio State College, I never again returned to my father's home, one reason being that he married again and we never were friends, and second, I married after I finished college and headed for the Alaskan frontier, where I broke American quarterhorses, which had become my love. The marriage ended in divorce, and I left Alaska, arriving in California in the fantastic days of the big studios, high-budget pictures, contract players, lavish soirees, and all the fanfare that made up the Golden Era. I acquired a quarterhorse and became acquainted with movie stars, and through Republic Studios I was put under contract as a starlet. After all, I could ride a horse in those low-budget B pictures I was assigned to. I made $150 per week — a bundle in those days. I never wanted to be an actress — I'd never be a good one — so I never made any waves, just kept out of sight and rode my horses.

My introduction to a tiger came at a Sunday rodeo in 1972. The man producing it asked, "How about appearing with G.T. (Great Tiger)? You're small and will make a good showing. She won't hurt you." My face must have worn a mask of unaccountable shock as I replied, "Me, appear with a tiger? Not on your life." I knew nothing about big cats. All I'd seen was a man in a circus in a cage full of them, equipped with a pistol, a whip, and a chair. But after a rundown on G.T.'s lifestyle of domestication, I consented, and rode around in the back of a convertible with her. After they parked the convertible, many spectators from the stands strode over and asked ridiculous questions. The man answered to everyone, "She eats people." I asked, "Why do you say this?" He replied, "After a while these questions are a pure pain in the ass. You mustn't be too hard on me. I discovered early that lying was the only road to survival with G.T. I have to protect my cat by joking. Dee, you will understand if you ever own a big cat." But I never intended to own a cat, big or small. Why in God's name would anyone want to own a tiger? It was the farthest thing from my mind. I had other interests; in fact, I had a plane ticket to Santiago, Chile, as it was ski season down there.

But I was somehow drawn to G.T. Was it her size, power, or her great gentleness that attracted me? Or was it because her handler had mentioned that a tiger could be a repellent against men, and in the recesses of my mind I knew I wanted to be free of any emotional entanglement with them. And

as I daydreamed on the plane back from South America, I knew I would see G.T. soon.

In fact, on the initial visit to the compound, little did I realize it would be the beginning of a brand-new chapter in my life. Her owner carried a squirming bundle of thick, long, yellow baby fur into the ranch house and put the petrified eight-week-old tigress cub on the floor beside me. She looked up at me as if I were the Grim Reaper. I stared down at the furry baby who was being sentenced to life in a cage. She had been declawed but she could bite, and bite me she did. Then the trainer showed me the correct way to hold her by placing his hand between her hind legs and under her stomach so she straddled his arm, and he held her chin in his palm. He tried to explain that tigers are loners and do not run in prides like lions. And this year he was ass-deep in tiger cubs. "This one is going to a circus, have to get rid of a few." Then he said, "Dee, how about playing foster mother to a cub?" In an insane decision I blurted out, "I have to have this cub."

I wanted to save this innocent angel. For some time I had needed an objective, a goal; this could be my chance for doing something for someone who would appreciate my efforts. And I knew I didn't want her to be a circus cat or caged. Ideally, what I wanted for this beautiful cat was to let her grow as unencumbered and free as the confines of society could offer. I just wanted to live with her and let her be as much a tiger, and me as much of a woman, as our union would allow. However long it would be, it would be sweet.

The owner prepared an agreement that said I would pay all expenses and agree to return her to the compound upon request. The realization that one day I might have to give her up hadn't registered in my mind. The panic of my insane decision to foster-mother the cub didn't set in until I arrived home, but as I looked at her lying on my bed I rationalized that at least I wasn't guilty of the sin of inflicting man's will upon another species strictly for personal pleasure or profit. My bedroom was her nursery, and I soon placed her litter box in the bathtub to prevent the litter from traveling. I started her on formula of baby cereal, canned beef, lamb, cottage cheese, and vitamins with lots of cod liver oil twice a day. Her prize possession was her teddy bear; she never tore into it, just licked it with her coarse, cowlike tongue. She roamed around the house freely, with a bell on her collar as my clue to her whereabouts. I learned to keep a close eye on her or trouble would follow. One day she disappeared into the bathroom. I arrived to find her polishing off my soap as if it were baked Alaska. There was no sign of my sponge, but from the fragments on her whiskers, I could see it had been an appetizer. I told myself, "Don't panic. After all, you can just phone the vet, and he'll provide a stomach pump in short order." Just ten minutes went by, and she threw up the soap and then the sponge. But I rarely left her when she was a cub. If I did, I hired a little ten-year-old for kitten-sitting, and I gave her the following rule: in case of fire, take the cub and leave.

In December of 1972 I received a phone call from B.C.'s owner's wife — he was dying of cancer and she wanted the cub back. They were selling all the animals, some to a circus in Mexico and Brazil. So now reality set in, and a circus was going to be her home after all. I could picture her staring into my face with those amber eyes and thought, "I mustn't let the charm of those walnut-sized eyes and her beautiful coat go to my head — after all, I didn't want a tigress or to get stuck with a 'big cat.' " But when I saw the owner's wife, I didn't muck about with words. I blurted out, "How much do you want for B.C.?" She looked surprised and said, "Dee, that cub is everything you don't need." So I asked again, "How much for everything I don't need?"

She came up with a figure that sounded something like two years' loan payments on my beautiful home. So I haggled like an outraged Frenchwoman at the flea market and we were in deadlock, but I had no choice; I didn't want to pay that price. The woman said she could ask any price for a rare Sumatran tigress, for she only knew of seven in the country. But finally we agreed and I then owned a "big cat." I had a lot of mixed feelings as I climbed into bed that night and began to soliloquize: What was B.C. going to contribute to my life? She was a living, breathing piece of nuisance especially designed to leave tiny scars on my arms from her small teeth, rob me of my sleep, destroy my furniture, wearing apparel, and peace of mind, and at the rate she was going, she cost me more to run than my new Cadillac — but she needed me, and now I was all wrapped up with a tigress.

Now I realized we must find a new place to live and sell my beautiful home, for bureaucracy was raising its ugly head: no tigers were allowed in residential zoning. So everywhere I went B.C. was sure to go, like Mary and her lamb, for L.A. Animal Control would have grabbed her if I left her. I looked and looked through zoning books. Animal regulations showed no mercy or exceptions regarding my roommate and companion, whose only sin was being born with stripes and being labeled a tiger. I went through very hard times of getting properties and having to leave them because of problems with permits.

I became reconciled to certain things as unavoidable as chuckholes in my road of living with a tiger. Trying to earn a living was one of the worst. With B.C. having to tag along everywhere I went, she prevented me from pursuing my vocation as a commercial designer. I began to feel like an American Indian squaw with a papoose strapped on my back. But a smooth area was our excursions in job hunting for B.C. We visited all the studios and production companies. She became an instant celebrity with her good manners. I hadn't planned on her turning into a regular Joan Crawford, but she even made a television pilot, though it was never sold. She made some commercials, and when we traveled she was always welcome in any motel.

But for the past eleven years my story has been one of constant struggle and determination to keep B.C.'s domesticated lifestyle, to which she has been imprinted since she was a cub. The most common question people ask is, "Aren't you afraid she'll kill you?" I try not to flare up or become defensive, but it's hard not to. I would love to say "I understand your blind fear, you're entitled to your point of view . . . yes" and go about my business of B.C. and me, but nobody is really going to understand no matter what I say. Even listening to myself, I don't understand. As the years have piled up, they sadly drew me a clear picture that indicated it would be B.C. and me against everything. We were somehow unnatural, and because of this, dangerous. Before I had B.C., if I had seen a lady living with a tiger, I too might have felt ill at ease.

She weighs 350 pounds now. I feed her in the morning and then again at night. I usually give her chicken necks in the morning, and at night I give her canned tuna, fresh liver, or kidneys. She eats about 5 pounds of food a day. In the summer she eats less. I give her oil and vitamins in her food. Most people just throw the same food in for them all the time. Since she's just a cat, I give her a pan of milk too. I also give her tapioca pudding, cottage cheese, and hard-boiled eggs. And her favorite is ice cream. I never bathe her, but I put a foam cleaner on her and then wipe her with a towel. She doesn't have fleas.

At night she comes inside. I used to let her in before I too[k a b]ath, but she chews on my bed too much when I'm not around. I lik[e to sp]end my evenings with her. I talk to her. I just really love her, and I [get he]r kisses. When she doesn't want to be bothered and she's not in a lov[ing moo]d, she'll just push me away with her paw. I enjoy B.C., and there is a [m]ore peace with her than there would be with a human being. You do[n't] have to be anything but yourself.

When she lies on the bed with me, she doesn't like me to get up and leave. If she's been sleeping for hours and something's wrong, like after she's been asleep and it's two o'clock in the morning and I forgot to turn the water off, I can get up and go out and turn the water off. But if I get up and she's only been asleep about an hour, it makes her mad and she'll chew on my bed. I got mad at her the night before last and I kicked her out and told her to sleep in her own bed. But about five or ten minutes later she was right back sleeping with me. She knows what you're saying to her, but she acts exactly like any other cat. She's no different from a small cat, and will only pay attention if she wants to.

Sometimes when we're asleep, she'll pull me over to her. In wintertime she will. And I just cuddle up to her. Some nights she'll turn over with her paws the other way and then I'll cuddle up to her and put my arm around her. They're supposed to be solitary animals, but we have a sort of unique relationship. I don't think there's another one like it.

I have a will made out. I would never let anybody put this cat in a cage. If anything happened to me, I would never let anybody have that cat. Never. I'd never let them mistreat her like I see other cats mistreated. I think this is the only tiger that gets a fair shake in this concrete jungle called civilization.

Sometimes, even to this day, I tend to forget that everyone does not have a tiger in their house. I have to remind myself that B.C. is as rare as a blue diamond in the five and ten, an oddity and novelty in this age. I don't really think I did anything with my life until I got B.C. To me it's important that I gave one animal freedom that would otherwise have been just a circus cat and ended up in a cage. Maybe I'm not worth much to anybody, but I've done something. I can't save the world. I'm just lucky if I can keep this cat alive.

Who Could Love a Cockroach?

When I first met Geoff Alison, he was living on the third floor of an apartment building in a working-class neighborhood in Boston. The only furnishings in the apartment were a mattress, a metal plant stand filled with small pots of African desert plants, and a two-and-a-half-foot-tall metal wastebasket. The wastebasket was home to fifty Giant Madagascar Hissing cockroaches, some of them over three inches long. Geoff kept the top of the basket smeared with Vaseline to keep the cockroaches from crawling out. The bottom was filled with dry sand, which Geoff changed once a month. Pressed-paper egg carton tops served as cockroach nesting boxes. A small food dish held a mixture of granola, powdered bird vitamins, and apples. Geoff said that their favorite fruit was oranges and that they liked other water fruits like peaches, melons, or grapes.

Geoff, a graduate of Clark University with a major in psychology, has worked in the past for a desert plant wholesale business and has also propagated and sold rare plants himself. But he has frequently been unemployed and has always had to live on a small income. He is unmarried, and now lives alone in New York City. His brother and parents live in western Massachusetts.

GEOFF ALISON: My time began twenty-six years ago, outside of Watertown, Massachusetts. My parents were city kids, brought up in New York. My mother was terrifed of just about any animal, and my father didn't know anything either, but they were reasonably curious people. My father worked 169 for the phone company, and my mother was a music teacher. I was born at six and a half months, so they kept me in an incubator for a couple of months. They told my mother, "Look, this thing is not going to live. Don't even think about it." The incubator had an elevated oxygen level, and it led to complete damage of my retinas.

My brother has done a little research on the subject, and he said that at the time I was born, it was already known that oxygen damages the retina, and they should feel terrible. But I get the feeling that I came here to do certain things, and this is not true of everybody. This blindness is something that I carry, and other people carry other things.

I remember being in the incubator. First of all was the realization that I existed, then the awareness of being alone. I wanted another being to come to me, and I sent out a call into the cosmos. Then I remember a sensation of "density becoming less," which could be translated into a loss of heat and ultimately of life. It was a global sense of something becoming thinner and thinner, then just chaos, a vaguely unpleasant chaos. This memory is different from any of my later thought patterns, like the different little things I remember about being a baby or learning to walk.

This experience was relevant to animals because it made me know intuitively that there is a deeper level to things. I think everything is overlaid on the common denominator of a nonsensory hovering in existence, in which one barely gets a distinction between "self" and "around the self." So different abilities, sensory stuff, and intelligence are an overlay. It is consciousness that underlies everything on a life level. A second part, which I will probably never fully know, is whether being blind threw me into a balance that made these realizations more available to me.

From my earliest memories I was aware of spiritual things. When I was two and a half I would think, "Why am I here?" and I would try and encompass the universe and try to understand. I would get this rush and would go vertically, and my head energy would mushroom out. Then I would blank out and feel myself sinking to earth with normal consciousness. I would do this many times a day. I knew the plants and creatures knew about this, and I knew people did too, but it was behind a layer in them. It's like my sense of reality is different than other people's, and I am left feeling a kind of constant culture shock.

At about two I was beginning to explore outside. My parents encouraged me to do what I could in my own way. My brother, who is about two and a half years older, was the gatherer of information. We did a lot together. We were very good friends as kids. My interest started out in the backyard with simple privet hedges and Japanese beetles. My experience began with insects.

In the fifties, Japanese beetles were beginning to be a plague. I remember

people talking about them and hating them. I knew a beetle could fly and that they ate plants, but I really didn't know anything else about them. When my brother first handed me one, it wasn't anything like I thought it would be. When you consider how small my hands were then, in scale you'd now have a beetle about the size of a quarter — a monster beetle. I was completely fascinated. You'd find beetles with different tendencies. Sometimes they'd freeze up and clutch, but my first one just sort of walked. I was completely amazed by the shell, the legs, and how they worked. Then I let it fly off my hand, and I was totally mystified.

And of course in those days there were grasshoppers. Then there were small beetles that we called black beetles and gold beetles. And there were earthworms, which I thought were wonderful. There were a lot of lacewings, but they are pretty fragile and I couldn't do much with them.

I went to the Perkins School for the Blind from kindergarten to fourth grade. I remember that they took us on a trip to the Museum of Science, and they let me touch the mammals and birds. I flipped, I just flipped. That was when I learned what most of the wild things really look like.

When I was about five I began to enter pet stores, and went into the usual pet store phase. We kept little slider and map turtles in bowls, and horned toads. We did what the pet industry said, which was of course wrong, and they all died. These were the sacrificial animals. When I was eight we moved to Springfield, which was more rural, and I got mice and hamsters and raised cecropia caterpillars. God, they get to be about a max of 7 inches and really thick. They are very docile, and have tubercles that are set in rows, and each has a ball of little spinelets at the tip. All along, I was very much aware of each animal as a living entity. And I felt their misery when they felt horrible.

I can understand why people don't like insects. They are certainly very different from us, and they have caused a hell of a lot of trouble in certain human cultural phases, when people had to struggle for survival. There aren't too many kinds of situations when you can afford to spend time hanging out with a cockroach. It's not exactly what would be considered productive activity. And sometimes I wonder if the human genetic memory goes even deeper. Insects are the only kingdom mankind has not been able to bring to heel. I think people at some level know that.

Being blind made me connect to animals through touch, and it put me more at their level. The visual power that human beings have allows them to analyze and act on an animal quickly, and it puts the animal at quite a disadvantage. I am in much less of a power level; practically put, it's simply easier for them to run away. I also learned to pay attention to body cues of animals. There is a whole underlevel of response that all bodies go through, from sea cucumbers and fish on up — small responses that get tested out, discarded, or held in abeyance. By physical touch you can read that. For example, if you've got a frog in your hand and it's going to jump off, there will be a series of almost prejump jumps. They prime themselves. I would be able to put my hand gently as a half-shelter over the direction they were thinking of going and sort of cut them off at the pass.

In 1975, a friend gave me two pairs of Giant Madagascar Hissing cockroaches. They are about 3 inches long. One was a very old male, and he soon died of old age. When they do this, they go into a stage of noneating and nonmoving, and they seem to progress into a gradual euphoria or death state. They become relaxed and utterly, utterly unrelated to the world — just like a butterfly does after it breeds, has its eggs and caterpillars, and has no more function in terms of reproduction of the species. Then, until the frost, butterflies are released from any sense of urgency. They just explore and get blown around with the leaves. They become fearless; they just don't care. Dying cockroaches are like that. They become totally quiet, and eventually they will roll over on their back voluntarily. They let their legs go limp, and they lie there for several days before they are dead.

There was a point when I had forty adults and no kids. Now I have several females, many of which are pregnant. You can tell the males because they have horns and the females don't. The size of a baby is about ¼ inch. They're all hornless when they're little kids. Only gradually do they develop, and they shed their plate as they grow. Laying the egg case is a pretty intense thing. The pregnant females are massively fat. The laying is obviously a very painful process. They tremble; they exert. They do muscle contractions, and it takes a while. The egg case is extruded and then resucked up into the brood pouch. I guess when the mother begins to feel the babies hatching, she extrudes them. When there isn't enough humidity for her to get the egg case back into the brood pouch, it jams. It's horrible. I've tried everything to save the animals. Sometimes I stroke the last part of the belly to get her to relax. If I can't remove the case, she'll die. Also, on a couple of occasions I have had to help them shed, but usually when they have shedding trouble it's bad. They go through at least half a dozen sheddings as they increase in size.

Being with them is a reaffirmation because for them every step, every move of the antennae, is a mind blast. They have a complete amazement at everything. Out in Madagascar they would be utterly still at daytime and hide under things to avoid birds. At night they would be up and about, eating, and with males climbing on the highest thing they could find and hissing. That calls any females to the area, and the males will have whatever territorial disputes they want. They have an organization. The duties of the leader are essentially guard duty. He will climb on top of the box and watch. The males choose a leader by pushing each other with their horns. The one that can flip the other becomes the undisputed leader. A few days before a leader is to die, he is accompanied everywhere by a very strong other one. They curl up together, they walk together, and it's usually that male that takes over.

Early on, I did an experiment by picking up their piece of bark and scattering them around the aquarium. The lead male and one of the females patrolled back and forth to meet all the babies and walked them back to the bark I had moved. They would not go under the bark until all the little ones were back.

When you first meet a cockroach, their response level is less than when they get to know you. They, like many small creatures, have a real need to feel surrounded, and try to get an extreme union at a body level with the space they are touching. So I begin by picking them up in a way that doesn't rip them off what they are holding. Then I try to move my body immediately to conform with how they are moving, so that my hand instantly begins to react. If they want to roll forward to protect their belly, I'll try to give them that shape, and as they begin to relax, I will begin to flatten out the part of my hand they are sitting on. Then I will partially cover their body so they feel that security, but I usually leave their head uncovered because they have a fear of having their antennae damaged. No matter how they were feeling when you picked them up (maybe you woke them up or they didn't feel like being picked up), there comes a point of relaxing and exploring. The body is less hunched; their weight almost becomes less. It's a weird sensation, but their legs have an almost springlike action, and when they relax them, their body almost bounces a little, as opposed to being completely jammed up with all the muscles contracting with fear.

The phase that comes after that is, they become completely relaxed. You can literally feel the spaces between the plates of their armor relax. Then I usually put my finger around them, and they'll eventually do a lot of touching with their antennae and front legs, and pretasting sorts of motions with their jaws. It doesn't hurt. They'll begin to sort of ooze in a direction, then get up on their legs and walk along one foot at a time, very slowly and delicately. The weight shifts slightly and the feet vibrate strangely as the leg tenses. Somehow the foot begins to detach at the same time it is holding on. This must be at a very micro level. Some get into enjoying doing acrobatics around my fingers. They have a great time going under, over, and around.

Each cockroach is an individual, having qualitative and quantitative ways of manifesting different responses. Given the same situation, one insect will have a more intense period of emoting or showing fear, or you may have one that's extremely curious or has an incredible memory. One, for example, can remember where the fingers are very quickly and not get lost when they change positions. I would say that cockroaches are capable of doing a lot of learning and a lot of decision making on an emotional and analytic level. As a life form, I'd say they are up there. They even seem to become very intimate, with definite bonds of friendship, and it isn't always between opposite sexes.

One time I was away for several days, and the person caring for them watered the sand at the bottom of their container. This was a terrible mistake. They are creatures of the semi-arid desert. Before I returned, a mold occurred, and all the adult males died. And the mold attacked the feet of the females and made them fall off. I just prayed that among the remaining young kids there would be juvenile males. During this period, one female assumed the assertive male role and would sit on top of the box. When the males grew up, she was sort of eclipsed, but the personal experience of doing all the things she had made her different. She remained very curious. One old female had no feet left at all. She had stumps, and of course for her to get anywhere took monumental effort. It would take her three minutes to get across the wastebasket. I took care of her. Whenever I'd bring new food, I'd have to pick it up and put it next to her. The others gave her a place to sleep next to the doorway. They were always careful of her, and they didn't sleep on her. So she developed total trust, and I would give her the best of everything. There was a sense of calmness and a certain kind of shared mutuality. She rose above her essential instinctual responses.

Between anything that lives there can be a psychic bond, and there was with her. We could communicate telepathically. She would communicate concrete things, such as she was glad of having something done, and nonconcrete things — that she liked to be with me and being very aware that I responded to her situation. When the time came for her to go into her euphoria, she just vibed me. I came in and she was on her back. I picked her up, and she said, "Well, the time of death has come." This was expressed in pitch, because when you die you begin to express a tone like a pitch rising in volume. She said, "Before it blanks me out, before I become so totally involved in it, I want to say goodbye. It's been nice and good being together." That was the last communication. She became limp, and several days later she died. She had been from the first litter, and died at two years.

At another extreme there was an incredible, strong, aggressive male that didn't like the idea of anything happening he didn't control. He was just outraged when I would pick him up, and he would actually butt my hand with his horns. When they push forward, they can exert a lot of pressure. And he would hiss. When they are afraid, they will hiss to where they are completely flat, and the hiss will have a shake to it. But when they are mad it is shorter, sort of like a burst, because they keep themselves puffed up. I could always recognize him. His thoracic horns were larger than those of the other males, and they were grooved and sculptured. He was beautiful.

I've had a lot of different insects, and I've had amazing experiences with almost every kind. I had one long-horned beetle that was pinned on her back. I heard her struggle, and I knew she was in trouble. They're wood-boring beetles and they can be hand-boring beetles too, but I decided to risk it. I slowly lowered my finger until I was within grasp of her feet. She grabbed on and pulled herself up. Then I turned my hand over, and she all of a sudden realized she was in my hand. She hunched down and opened her jaws. Then she stopped, thought about biting me, and relaxed. From that day on I could pick her up.

An insect that's outrageous is a monarch butterfly caterpillar, the younger the better. You have to give them fresh food twice a day. They sleep at night. You have to be very polite with them, but when they're not eating or drinking you can hold them. First they're likely to hunch up, but quickly they learn to relax. And they'll make a chrysalis and come out in fourteen days as a butterfly and follow you around. But I had one whose wings never dried right, and he couldn't fly. He had to live with me. He was the greatest, most friendly creature.

That monarch was the most outstanding insect I've ever had. He would go everywhere with me in the two months I had him. He learned to use his wings as a sounding box, and he created a language. Anytime I would hear him walking or fluttering I would respond, and would put honey water on paper to feed him, and that budded into communication.

In the morning at about a quarter to eight he would wake me up by doing a rhythmic thing with his wings. I would get up and feed him. When he was done eating, he would stand up on his hind legs, and I would put my finger down and he would let himself be lifted up. Then he usually wanted to try to fly, so I would raise my hand as high as I could toward the ceiling and he would jump into the air. Then he would flutter to the ground like a winged seed. After several minutes he would get frustrated and wouldn't want to do it anymore. Then he usually wanted to sit up on the curtain pole, in the sun. They love the sun. He'd make a motion with his wings. The way I learned was, if I didn't do the right thing, he'd keep doing whatever he was doing until I got it right. I would try food, water, holding him, putting him in the sun, letting him fly. There were times when he wanted things that just couldn't be expressed. He was like a baby, and he would get frustrated. Usually the main part of the day he rode on my shoulder. I had to be careful outdoors, because sometimes he'd get the notion to flutter and almost got stepped on a few times.

What finally happened to him is really amazing and really sad at the same time. As his migratory urge came up, he wanted to fly more and more. I was awakened one morning by a new vibe that didn't sound good. He was

perched on the top of his tray, and in his frenzy to fly he had damaged himself. He had only about half a wing left. There was one foot left, and there was blood oozing from many abrasions on his abdomen and underside. He was dazed, frightened, and in pain. I held him in my hands, and he calmed down and pulled himself up on his stumps and just waited. I realized what I was going to have to do, and I vibed it to him and he accepted it. I went outside and put him on a rock, and I put my hands around him and waited until he put himself into a position of readiness. I took my hands away and vibed him, "Well, this is the time," and I smashed him with a rock. Instant death. Death is really a strange thing. I hope I can die as nobly.

Well, I've always known what the purpose of my life is, but I still haven't figured out all about the method. The purpose is to help forge a means of positive communication between living things. Something that mankind has to learn to survive is that the connection of two living things is a microcosmic expression of the beauty of the universe.

Thirty Tiny Terriers

Irmgard Taylor is an expert breeder of Yorkshire terriers. She and her husband, Harold, have lived in the South Bay area of San Diego in the same fifty-year-old Spanish-style house for over nineteen years. A spacious 50-by-150-foot front yard and a backyard the same size give ample room for Ms. Taylor's plants and animals. Her mother, Fina Laub, eighty-three, lives with them. The Taylors have grown children with families of their own, and nine grandchildren. With her children grown, and her husband on the road with his trucking business much of the time, the companionship of her animals is important to Ms. Taylor.

Her little terriers follow her everywhere, forming a veritable moving carpet of blue-black and red-gold. Each animal has a miniature topknot, secured by a rubber band, to prevent its hair from falling into its eyes. Their noteworthy uniformity and small size are the proud result of careful breeding.

The babies are grouped in padded playpens, and are fed goat's milk, baby food lamb, and beef. A covered balcony was constructed at the back of the house so the Yorkies can get fresh air or sun when they don't accompany their mistress outside. At night, each dog goes into its own white molded-plastic air kennel in which a bedtime food treat is waiting.

IRMGARD TAYLOR: I just keep as many dogs as I can take care of. I have thirteen now, but I have also had up to thirty.

I was born in Heidelberg, Germany, and was nine years old when we moved to Ludwigshafen, on the Rhine River. I was then in the war. Oh yes,

I spent the whole war in Germany. I got bombed out three times, but we were lucky, and no one in the family was killed except uncles who were soldiers. We were bombed every night. In fact, my one daughter was born in an attack. I was laying there to have my baby and the windows broke, and people were running around because they had to take all the patients down to the cellar. I was laying there having my baby almost by myself.

I always had dogs. When I was about three years old I had a black and white fox terrier, Fritzy. My whole family was animal-loving. I also had a big boxer. My parents loved animals, but my mother's father was the biggest dog lover. He would not eat without feeding his dogs. When he was sitting there eating his black bread and piece of cheese, the dog got the same thing. My mother says I take after him. I can tell a story that people wouldn't believe. When the war was going on, my grandfather had a mixed dog, Maxy, and he was frightened from the bombing. So my grandfather was giving my cousin the dog, and that was far from our town. And through the war and bombing, Maxy went all the way home to my grandfather. He was so skinny, and his feet were bleeding, and a neighbor came and knocked on the door and said, "I think your dog is outside." And he said, "No, my dog is far away." "Well," he said, "it looks like your dog." And it was his dog.

Of course, when we went in the bunkers, we couldn't take the dogs there. Well, I sneaked mine in. I put the fox terrier in a blanket like a baby. But one day he just disappeared, and I hate to say it, because I don't know if it's true or not, but they say the Russian prisoners we had there ate him. He was very old by then.

My husband died in the war. He was in the Snow Alps, in France. He was walking in front of the company and he had already a stiff right arm, but they said he didn't have to shoot or anything, he just had to command. He was killed by a sniper. He had loved the dogs too.

I remarried in Germany. My second husband tolerated dogs because I liked them. We moved to London in between our traveling from Germany to Austria, and stayed from 1957 until 1959. I went to dog shows and saw Yorkshire terriers for the first time. I thought, "They are little but very spunky. They are afraid of nothing, and they never lose any hairs." When I came to America, I bought one from a lady from England and paid $500. It took my husband two years to find out how much I paid for the dog!

Then I really fell in love with Yorkies, but I didn't so much want to be a breeder. I am a dog lover, and there is a difference. If I were a breeder, I would just put away my dogs and feed and breed them. But my dogs are my life. Also, I would not breed a kind of dog if I thought it would wind up in the animal shelter. I have talked to people there, and a man told me, "If I have one maybe in three years, people are already waiting when I come."

179

Of course, my kids are all married now, so these little dogs are just like my kids. My son says when he comes in the house if he wants attention he has to bark, but he likes dogs. One daughter has cats, chickens, and dogs. The other has a Yorkshire. My husband is on the road all the time because he has a trucking business, so I have my animals to give my love and attention to.

I get up in the morning, and the first thing I do is put my automatic coffee maker on. Then I let my dogs out in the fresh air. Then I feed them. I read and make embroidery. When I have puppies I play soothing music. When I go swimming in the pool the dogs run back and forth, exercising. When I stop they kiss me. If I am feeling blue I take a shower, put on some cologne from Germany, make a cup of coffee, put on music, and the dogs sit with me. In the late afternoon they go out for two hours and sit under the orange trees. Then at five or six at night I say, "Come on kids, let's go," and they go into the airline flight kennels where most of them sleep. Each night I put in a treat for them, and they like going in there. Liza sleeps with me because she is a watch dog. She will let me know if somebody is coming. Little Botzel sleeps with me too.

I had one that lived to fifteen years. She had a little pillow by my bed. One morning just recently I went up and I said, "Rumpy" — her name was Rumpelstiltskin — "what's the matter with you? Don't you want to get up?" And she had just died in her sleep.

Botzel, which means "little" in German, is the smallest one. She has a beautiful coat, but she is not strong and has a breathing problem. A lot of people have offered $500 for her, but I wouldn't sell her because I know just how to take care of her. Oh God, when she was born she was just a bit bigger than my little finger. She had no hair. She was just like an eel. The others were twice as big and of course they had hair. Even her mouth was not really formed. I had to feed her all day and all night for six weeks, every hour. First I gave her a solution of Karo syrup, water, salt, and salt substitute. Then goat's milk diluted a little bit. When I slept I put my hand on her, and every time I woke up I was thinking, "Is she still alive?" I really believe the dog is still alive because she wanted to live and was just fighting and fighting. She had bad days when I thought, "Oh my God, I don't know if she makes today." For six weeks I didn't go anywhere, and I carried her in a little towel that I warmed up in the oven. Then for months I carried her around with me in my pocket and kept giving her drops of milk or vitamins. Now she weighs 2 pounds and is the most feisty one in the bunch. I respect her because she fought for her life.

If I know a dog will have puppies, I put her away from the other ones about a week before. It would be a terrible stress for a dog to have puppies being with other dogs. I put her beside my bed, and I keep my hand there all the time to see what's going on. It's funny the way my mind is trained. If a dog is going to have puppies, I can wake up anytime I want.

I would not raise dogs unless they went to a good home. I am very particular where I sell my dogs. I had one man coming, and he asked me over the phone if I had a Yorkshire for sale for his wife. I said, "Well, come and look." He said, "How much does it cost?" I said, "Three hundred and fifty dollars for the female." He said, "Okay, I'll pay you with a check." Then he came with two little children, and I said, "Oh, wait a minute, sir, are they your children? I am sorry to say, but I cannot sell this little female." He said, "Well, my wife wants a Yorkshire terrier." I said, "Yes, but those kids will want to play with the dog. Why don't you go to the Humane Society and get yourself a dog the kids can play with, and then when they go to school you can get your wife one." The man said, "What do you care? Here is your check." I said I was sorry. The man was mad, my God. I wouldn't blame the kids, but they could have squashed the little dog.

If an old lady comes and she wants a dog and she doesn't have much money and I think my dog is getting a good home, I make an exception because I know my dog will be the baby there. Before I sell a puppy, I give them shots and try to paper- and leash-train them. Then the people are happy, my dog is happy, and then I am happy.

I try to better the breeding. One thing I think is very important is that you pay attention to the personality. A lot of breeders don't care. I look for a dog that says, "Hey, I am here too," but that doesn't fight or bark. I like them to stand nice and proud. I like nice color, gold and blue. Nice ears and a nice body. Straight back. I like a long neck on a Yorkie. I like the tiny face. Some are born to show off. When they walk in the ring they look at the judge, and it's like they say, "You don't have to look at those others. Look at me." By six months I can usually tell if it is a show dog.

I keep books on each dog when they are shown. My dogs win their championships quickly, and it makes me proud. Some people show a dog for years just to add up the points for champion. And my dogs win with many different judges. Some have both their American and Mexican championships. After they win they come home, and I cut off their hair so they can run around. I want them to have freedom. Some people just groom them and put them back in cages. They tell me that they even have cages where they can pee in there and it just goes down so you never have to take them out. I think this is something awful. There are a lot of people that are selling Yorkies that don't even look like a Yorkie, and they get $300, too. This is a shame, and it ruins the breed. I don't really make money on the dogs. I put it back into their care. I also pay my handler, whom I have had for five or six years. The dogs like her.

I am sixty-one now, and I don't know how long I can continue with the dogs, even though I feel thirty or thirty-five. But my dogs are happy, and they will stay with me until they die. I don't care if they get old or if I don't breed them. I just keep them.

I can sit right here and thank God. I went through the war eating very little because there wasn't enough to go around, while this and that were being destroyed around me, and now I am living in Paradise. I think I made my own life, and I think I am lucky. My kids are good citizens, and I always have the little dogs around. I think I understand them very well and they understand me very well. With dogs like this, it doesn't matter if you have a bad day or a good day, they love you. I feel I am not alone.

From Ants to Pheromones, to Sociobiology, to Biophilia, to Consciousness, and Back to Ants

Edward O. Wilson is the Frank B. Baird Jr. Professor of Science at Harvard University and curator of entomology at its Museum of Comparative Zoology. In the late 1950s he helped to revolutionize notions of animal communication by research on chemical transmitters called pheromones. Then he studied biogeography and conservation ecology. In 1971 his *Insect Societies* was hailed as a monumental synthesis of knowledge of social insects. *Sociobiology: A New Synthesis,* published in 1975, was a compilation of his theories on the behavior of human beings and other animals. Its central theme is that patterns of social behavior, like physical traits, can be inborn. In 1979 Dr. Wilson won the Pulitzer Prize in nonfiction for his book *On Human Nature.* Two years earlier he had received the National Medal of Science from President Carter.

His laboratory is half of the renovated fourth floor of the laboratory annex of the venerable Museum of Comparative Zoology. The other half of the floor is used by a fellow research worker in social insects. The corridors are decorated with art from Dr. Wilson's books and by dramatic photographs of ants enlarged drastically by scanning electronmicroscopy.

Dr. Wilson's area includes graduate student offices, the office of his assistant, his research laboratories, and his own office, where he has his two colonies of leaf-cutting ants. Each colony is kept in an artificial nest composed of about ten clear plastic corked chambers, 5¼ by 7½ by 3¼ inches, connected by clear plastic tubes about ¾ inch in diameter. The chambers contain the fungus gardens of the ants. A long tube leads to the foraging arena, a container in which leaves used as a substrate for the fungus are left daily.

The professor's time is divided between his research and his teaching of Evolu-

tionary Biology, an introductory course in the core curriculum, and his advanced seminar in the spring, the Biology of Social Insects. He commutes about 10 miles from his home, where he lives with his wife, Irene, and his daughter, Cathy.

EDWARD O. WILSON: I am an Alabamian, born in Birmingham. You can tell a native Alabamian, incidentally, because they say "Alabamian," not "Alabaman." I was an only child. My father was an auditor — a civil servant — primarily with the Rural Electrification Administration. My mother was in public relations and management in the air force. They divorced when I was about seven. No one in the family had an interest like I did in natural history.

I grew up in Alabama and northern Florida and had the advantage of having ready access to the woods. I like the feeling of submerging myself in the surroundings of a forest. I can scoop up a couple of handfuls of earth, and know that the organisms present have more complexity among them than the entire surface of the planet Mars. I could spend a lifetime exploring what I hold in my two hands.

Human beings are fascinated by animals and plants, but they tend to grow up conscious only of the larger life forms: lions, tigers, horses, dogs, orchids, oak trees, and the like. But if they are able to change their scale of magnification even slightly to look at organisms at the level of millimeters, then the possibilities of the same kinds of excitement and serene pleasure that these objects give is both immediately available and infinite in prospect.

To me, a dragonfly can be just as interesting as a hawk, and an ant colony just as interesting as a wolf pack. I felt this right from childhood. At nine or ten I started collecting butterflies. They say El Greco didn't just have a magnificent artistic vision but may have suffered from astigmatism. My real reason for an interest in tiny creatures had a similar origin. I have the use of only one eye. I injured my right eye when I was seven years old in a fishing accident. Though I am only monocular, I have exceptionally acute vision, twenty-ten, in the good eye. Also, I am a little deaf in the upper registers. I was never able, even as a child, to hear the trilling of a lot of different kinds of birds, and so have never been a very successful birdwatcher. It may well be that these physiological limitations are the reason why, when I walk into a forest and start to examine nature, I scale down much more readily than most people and in a highly focused manner.

All my life I have been able to see more than most people when looking at a little flower or an ant. I didn't realize that I was in any way different until I was in my forties. I always assumed that everyone saw the world as I did, and that what they saw was fine and detailed. I couldn't understand why their attention wasn't readily caught by little flies and ants. That peculiarity 184 in vision may have contributed to a relishing of the peace given by these microscopic moments and the fact that I can find a very high degree of order and pleasing detail in humble surroundings.

I have always been interested in animal behavior, but during the fifties and sixties most of my attention was directed to taxonomy, ecology, and geography. When I was a graduate student at Harvard I did a lot of work with biogeography, the distribution of animals around the world. There is a grandeur in the way organisms have spread across the face of the earth. But a whole new interest began when, in 1953, Konrad Lorenz came to Harvard and gave one lecture on his new understanding of animal behavior, particularly of ducks. It became clear that ducks were seeing each other in a very different way than people had realized. They were probably paying attention only to a few key stimuli: whether the other individual had a red breast, whether it was making a certain sound, and so forth. More importantly, these astonishingly simple stimuli evoke very complicated behavior.

On that day I said to myself, if the birds are communicating by visual and auditory releasers, the ants and other social insects must be communicating by chemicals. I didn't follow the realization up right away. At the time I had a great deal more biogeographical work to do in the field. But I held the idea in the back of my mind, and when I turned more to laboratory work at Harvard in 1958, I decided that I would set out to see what sort of chemical signals I could identify in ants. One of the first things I looked at was the trail system.

Up to that time, we knew that ants had chemical trails, but they had never been characterized, and no one knew where they came from. It was thought that they only served as orientation devices, like slashes on trees to lead you through the forest. In other words, the ants found the smell and merely followed it. I decided I would try to find the glandular source for the trail substance in fire ants. So I dissected the abdomens of fire ants and made preparations from each gland in turn and drew an artifical trail to see if ants would follow it. There was no effect until I came to the little Dufour's gland, an almost invisible structure in the rear of the abdominal cavity. When I made a smear of that, literally hundreds of ants not only followed it but came tumbling out of the nest in high excitement.

I couldn't sleep that night. I kept thinking, "There's far more going on here than anyone realized." There is a specific substance in the trail; it is located and stored in one gland, and it causes a great deal more than just orientation. It is a specific and powerful signal from one ant to the others. Then I set out to do similar studies on other glands and other behaviors. I postulated very early that ants communicate primarily by chemical signals called pheromones, each of which comes from a different gland. We communicate by sound and by vision primarily, but the ants obviously do not. The fact that they communicate mostly by chemicals and live in a world of   185 pheromones is why they are so alien to us. We study them by vision and sound. If they studied us, they would be equally puzzled because they would try to study us by the smells that we were emitting. An ant scientist would approach people in conversation, and she would have to conclude that they were just sitting around with no communication because she would be running back and forth testing various odors and would find no trail substances recognition odors.

My work on sociobiology ultimately stemmed from these thoughts and observations. Its contribution is that it permits a much deeper, and in my view more interesting, explanation of all social behavior. Most people who have an interest in animals of any kind are fascinated most of all by their social behavior and organization. Sociobiology directly addresses the great spectacles of wildlife of the world, such as bird flocks, antelope herds, and lion prides as well as the less dramatic but appealing behaviors of sexual bonding and parental care. These phenomena have been given a new meaning and interest by the kind of evolutionary analysis introduced by the sociobiologists. The basic assumptions are that many behavioral traits are genetic, therefore inborn, and that social behavior goes through the same process of evolutionary selection as do the physical traits studied by Darwin.

For many people it is mind-boggling to consider that our perceptions such as "Sugar is sweet," "Women are attractive to men," and "The home soil is sacred" should be programmed in our brain. In fact, what is programmed is a strong tendency to learn one thing over another. Phobias are a dramatic example. They are directed at the ancient enemies of mankind: height, close spaces, snakes, rodents and a few other threatening animals, and running water. Ironically, many of the phobic responses are relatively irrelevant to modern life. The truly dangerous objects that have evolved in the last hundred years or so, such as electric sockets, automobiles, knives, and guns, are not the object of phobias. This is striking evidence that our brains engage in what the psychologists call prepared learning. The mind follows circuits set in place; we are far more likely to get emotional pleasure or pain from certain responses than from others.

Sociobiology has revealed why social behavior evolves in some animals and not in others, with special reference to the environments in which the species live. It has also gone on to explain the ultimate meaning of cooperation and altruism. In a few species, such as lions and dogs, a relatively high degree of cooperation among the adults occurs up to the point of self-sacrifice, whereas in most kinds of animals such behavior is relatively rare. It has been shown that cooperation and altruism promote the survival of kin even if the individual itself is hurt, so that the genes shared by the common descent are multiplied in the next generation. Also, up to the present, scientists have tended to reason from animals up to man; that is, to study animal behavior because we think it is so much simpler. But, in the studies of sociobiology, the possibility exists of going the other way. For example, if we know precisely how our color vision is organized at the cell level, we can make detailed comparisons with the brains and responses of the great apes and other Old World primates and infer how these animals see color as well. And at an even more complex level, we might also understand to some extent how these creatures feel pain and sexual desire.

It's entirely possible that an orangutan or langur has emotional responses, and possibly even conscious experiences, that are comparable to human beings'. We do have a common ancestry and certain outward similarities in mother-infant bonding, group affiliation, fear of strangers, and so forth. I believe it's not out of place for human observers studying primates to use a little bit of empathy in forming theory.

The biggest difference between human beings and animals, other than symbolic language, is our advanced form of self-awareness and the ability to reflect on the past and future. We carry the painful burden of being able to think about our own mortality, our declining health with age, and the content of a future world without us. Chimpanzees, so far as we know, do not think in this way. Higher animals have what can be termed free will, an ability to choose among perceived options. It's somewhat constricted in an ape and much more so in a dog, but there is enough consciousness and reflection to make a choice, and enough anguish or pleasure connected with it so that they may think about the outcome of alternatives. I believe they can be classified as having a mind.

But the concept can't be applied in any manner whatever to an ant. Insects have very little choice and almost no long-term memories. So one can hardly say that an ant has a mind, in the sense of having a consciousness and making a choice. It would be sterile anthropomorphism to speak of mind when you get down much below the mammals. I never empathize with ants. I recognize them for what they are. Their brains are just too small: a hundred thousand to a million neurons. We operate with a hundred billion. The difference is almost unimaginably great. So I look at ants very much as marvelous miniature robots.

The ants have a compelling interest for those who look at them closely. I started culturing leaf-cutting ants ten years ago because they are the ultimate social insects in terms of colony size and social complexity. The colonies at maturity have more than a million workers. It is believed the queens can live in excess of twenty years. The nests penetrate the soil in the tropics of Central and South America to a depth of 20 feet. (My largest colony was collected in Guyana.) The colonies are organized by a complicated system: the smallest workers raise a symbiotic fungus in the nest. They are a hundredth of the weight of the big soldiers that defend the nest against larger 187 animals. I felt that if I could understand their social system, I would be able to understand anything in the social insects.

In the laboratory, the leaf-cutter colonies are put in plastic boxes approximately the size of the chambers they normally dig. They thrive with a minimum of care. So even though they have one of the most complicated societies of any insect — or animal, for that matter — they are very easy to keep. I call my way of studying colony organization the pseudomutant method. I remove whole castes and reshape the colony the way a genetic mutation would if it eliminated the ability to produce foragers, soldiers, or some other social group. I study the effect of this pseudomutation on the colony, then restore the individuals and put the colony back to normal. I see the colony in this respect as a superorganism, the giant diffuse equivalent of a rabbit or guinea pig. Conventional biologists, working with more familiar animals, sometimes have to mutilate or kill their subjects. But with my pseudomutant superorganisms, no one is hurt. At the end of the day, the colony is restored to its whole condition.

I would say that a colony with sixty-five hundred workers requires no more than an average of about ten minutes a day for care. My assistant or I drop the leaves in, and that doesn't take any time at all. Once every couple of weeks, one needs to remove some old boxes full of refuse and clean out the foraging chamber. Furthermore, the ants aren't fussy; they accept a wide variety of leaves. Since they feed exclusively on the fungus, anything is fine that fungus can decompose.

In nature, leaf-cutters and many other kinds of ants put their dead in special refuse chambers. Oleic acid shows up in substantial amounts only in corpses. So the workers don't have to recognize that a sister has stopped moving or that it no longer has a head. They only have to smell oleic acid. In my earliest experiments with harvesting ants, I would put oleic acid on healthy, live workers. They were picked up by their nestmates with their legs kicking and dumped into the refuse pile. They would get up and clean themselves off, only to be dumped again. Eventually they managed to clean themselves enough to manage to rejoin the living.

I believe that we ought to consider the possibility that the human mind is assembled biologically in such a way that it has a natural affinity for other organisms. We form attachments to them; we raise them; we metamorphize them; we are intrinsically fascinated by them. A great deal has always been at stake in this symbiosis. For thousands of generations, human beings formed attachments to animals and plants and developed a detailed knowledge of them as a straightforward procedure for survival.

Biophilia is the term I use to label the deep and widespread tendency of people to affiliate with living organisms. The merit in exploring biophilia as a concept is the contribution that such understanding can make toward a lasting conservation ethic, one based on fundamental psychic need rather than on contrived economics or weak aesthetics.

In 1979, I asked ten Harvard professors what, in their opinion, was the worst thing that could happen in the 1980s. My answer was species extinction. Of course, nuclear war or economic collapse would be terrible, and yet, except for perhaps a global nuclear holocaust, those things can be repaired in a few generations. The one thing that will take millions of years to correct is the loss of genetic and species diversity. This, I believe, is the folly our descendants are least likely to forgive us. We would be depriving these future human beings of a large part of their planetary heritage, constricting them in the fulfillment of a profound inborn need.

What the Animals Tell Her

Beatrice Lydecker has titled two of her books *What the Animals Tell Me*, and *Stories the Animals Tell Me*. And she means what those titles suggest: she believes that she can communicate directly with animals, and they with her. Many other people share that belief, and they bring their animals to her for consultation. One of these people told me of the following experience, which confirmed for her the accuracy of Beatrice's readings: "I had taken my horse to the vet, and he was terribly upset and nervous when he got there. He was quivering all over. Weeks later Beatrice came to talk to him. I asked her if he was mad at me for taking him to the vet. She immediately said, 'Who was the man that went along?' I'd completely forgotten that my dad had driven the trailer, and I told her. And she said, 'Well, he blames your dad for the situation, not you.' My horse remembered what I had forgotten, and he told it to Beatrice."

My first meeting with Beatrice was in her house with its adjoining dog kennel in a commercial district of El Monte, California. She invited me to one of her readings, which was to take place later in the day at an elegant stable. There I encountered a most amazing sight. Almost fifty people, each accompanied by a horse, a dog, a cat in a carrier, or even a bird in a cage, were waiting patiently for her to tell them what their pets were thinking. The questions they asked were a poignant symbol of the communication barrier between people and animals: Did the animal like them? Could they make the animal happier? How could a particular behavior be dealt with?

When Beatrice "talks" with animals there is no trance, no show of concentration, no elaborate interaction. She begins speaking almost immediately, and states: "Your animal says . . ." She then says things which are surprising in their detail, and they are often later confirmed by the owner.

BEATRICE LYDECKER: I was born on Long Island and raised on a farm in Upstate New York, around a place called Salem, near Saratoga. There were five brothers and two sisters. We used to go to the horse races every Thursday night, the harness races at Saratoga.

My family was never into animals like I was. We had a dairy farm and horses. It was a living. The farm eventually failed because my parents weren't farmers; they were city people who wanted to get out of the city. They looked at animals as something that didn't have feelings or intelligence.

I grew up with goats, dogs, and horses as my companions. I remember standing and watching my dad drowning puppies because we didn't want more dogs, and I always thought that was normal. Dogs had puppies every six months and you just drowned them. It didn't bother me then because I had been taught to be insensitive. Animals were just animals. You killed them for food or they just existed. Sometimes now I think back on it, and pictures come to me of the ways the mothers felt: "Why are my babies in that water? What are you doing to my babies?" The pain they went through. What tremendous ignorance there was, but it gives me a little bit more tolerance for people who are raised with that kind of mentality.

When I was thirteen we lived in a haunted house. It really was haunted. Every family that lived there had a member of the family die. I know there were ghosts, 'cause I had experiences with them. My bedroom was right next to the attic, and every night I had to pull the covers over my head and go to sleep as fast as I could to forget it. Every night the attic doors would be shut and I would shove chairs and stuff up against them, and every morning everything was pushed back and the door would be open. One morning at two o'clock my mother died of a heart attack, the year we moved in. Then we left that house.

I went all the way through high school and then to an interdenominational Bible college in Columbia, South Carolina. I wanted to be a missionary to orphan kids. I was going to be sent to Cuba, but then Castro took over. I was twenty-one. Nine years passed, and I feel I ended up doing what God fully intended me to do right from the beginning. But I went through a lot of experiences. I worked in substitute teaching. I ended up getting married and divorced. I couldn't have children.

In 1969 I was going to a little prayer group when someone gave me a copy of J. Allen Boone's book *Kinship with All Life*. It was the first time I heard of the idea that there was such a thing as talking with an animal. Then one day about four months later, I walked up to a German shepherd that looked very much like one I had lost from a brain tumor. Suddenly I knew what the dog was thinking and feeling, and why. He was feeling very blue, neglected, and left out. His whole life was changing. I saw the owner later in the day and I said, "Your dog looks depressed." He said, "Yes, this is the first day

my dog has been by himself, and from now on he will be alone." So the communication with animals began all at once for me. And then it began to happen more and more.

I communicate with animals exactly the way one would visualize experience in a dream — I see, hear, and taste, except I'm wide awake. I've done it so long and so much now that it's like carrying on a normal conversation.

I can hear it in my head, but it isn't a sound you can hear with your senses. Sometimes I will talk to an animal from another country. The pictures of house, stable, barn, doghouse, will be the same even though the designs are slightly different. I can hear languages being spoken, but I can't really understand. Sometimes I can identify the country if I recognize the language. I can also feel what they feel. For example, when one horse told me her hip hurt and I touched it, I could vicariously feel a weakness all the way up in my own hip.

All young children have the ability to communicate nonverbally. When they go to school and learn more verbal skills, they lose the nonverbal communication. They also lose it because parents make them feel the nonverbal communication is a figment of their imagination. The kid will come in and say, "Mommy, the dog says he wants this," and the mother says, "Yeah, sure he does!" And so the child shuts off. By the time they get to second or third grade, the ability is pretty well gone. I don't even charge kids twelve years old and under for my class because they pick it up so fast. They're still close enough to it that they get it back. It drives the adults crazy; but the older we get, the harder it is to visualize.

There are thoughts going through animals' minds all the time, just like there are thoughts going through people's minds all the time. When I want to know something specific, I have to picture the animal in the circumstance. For example, if I want to know why a horse refused to jump in a horse show, I have to visualize the animal approaching a jump and stopping. In the picture I get back, the horse will not be in the picture. Instead, he will show me what he is seeing: the jump coming toward him. Suddenly I feel what he feels as he is approaching the jump. So first I am projecting a picture onto them, then I become them.

Sometimes there are things they don't want to talk about or they don't want you to know, and I keep forcing a picture on them until they give me a response. In most instances, they are happy to tell me. Since most animals have grown up with nobody understanding them, they don't really expect anyone to try to communicate with them. And what the animals say isn't always what people want to hear.

When people can communicate with animals, it changes their attitudes toward their pets, so the pet can lead a better life. I have taken dogs that people couldn't get near, and in a few days I'm loving them and petting them, 'cause they can tell me what's bothering them. Animals ask so little from us — just some exercise, some niceness, some love, and some play. They have a philosophy of not questioning why they're here, they just want to have a purpose. I've had many of them tell me, "I don't want to die, but I don't mind dying as long as it's for a reason. Just don't kill me to kill me." The animals that die in the pound have a really traumatic situation. They haven't been able to carry out the purpose of their lives.

Early on, I felt I had to know the truth about eating meat. I went up to a ranch where beef cattle are raised. When I passed I saw four or five in a bunch and felt them talking to me, and I said, "What's happening with you? Why are you up here, separated?" They said, "We're gonna go to the slaughterhouse. We're gonna be killed and eaten." And I said, "You realize that?" And they said, "Yes. We don't serve any other purpose, and it's all right with us." I turned to the owner and said, "Well, how come these cattle are up here?" And he said, "They're females who can't conceive; we have no purpose for them. We can't just let them live for years and years on the range. We raise them for money and beef. So these are going to be eaten." I was really kind of surprised, but it was a total acceptance, like it was their destiny in life and it was okay with them.

Then I went down to the slaughterhouse and asked the animals about their life up to that point. They said, "It's been wonderful. We've had a really good time. We live out in the pasture. We've had plenty of good food. They take care of us if we get sick."

So I went down to the kill floor, and I tell you that was traumatic for me. They were coming down the chute, and there was one steer standing on the side of the door and he was scared. His eyes were rolling and he was scrambling around. I said, "What are you afraid of?" He said, "I don't know. I hear the cow on the other side holler and I don't know what's gonna happen to me." I realized that he thought he was going to be abused. So I visualized for him what was going to happen: that they were going to kill him and it was going to be over quickly and he would be leaving. And he said, "Oh, is that all it is?" And he stood right there and calmed down. He said, "I'm not afraid of dying. I just didn't know what they were going to do to me before they killed me." And he walked calmly onto the floor. You see, animals accept death 'cause they know they're gonna live on, so they have no fear of losing their life. It's over like a snap of the fingers. They have this big, gunlike thing with a big spike nail in it. They cock it, place it right between the eyes, pull the trigger, and in seconds they're dead.

I was also there with the kosher beef. They tie the animals up, and it's only thirty seconds that the animal is in pain; then he goes unconscious because he's hanging upside down and all the organs in his body are pressing on his lungs. I watched them cut their throats and let myself feel what 193 they felt to know if they were in pain. And all they felt was a pressure as if you drew a line across your skin with your fingernail. They could feel the pressure of the blade, but there was no pain 'cause they were already almost dead from unconsciousness.

So it made me understand that they realized that if they weren't being raised to be eaten, most of them would never have had existence. Even when we human beings are born, we have no guarantee how long we're going to live. After I experienced that, I no longer had any qualms about eating meat.

When I started communicating with animals, friends got wind of it, and in about 1971 I had a newspaper interview. The reporter brought his dog and asked if he was intelligent, and I said, "Well, he is average." He got so mad that he wrote a sarcastic article about me. Meanwhile, so many friends were bringing me their animals that I didn't have time to work. So I decided that if I was paid it would be possible for me to do more. I charge $45 for an hour. If there are groups with a whole bunch of people, I charge $15 an animal and give them fifteen minutes, so it comes to $60 an hour. But it really saps me when I do a lot of them. You are constantly tuning in to a new animal. Once you are into their wavelength it is a little bit easier to communicate with them, but when you are changing animals every fifteen minutes it takes a tremendous amount of energy.

My ultimate goal is to do a television show because I feel I can reach more people that way. I would like to interview animals and get their opinions about specific circumstances, like dog shows and parades. I also breed my German shepherds and sell the purebred pups for $300. My own puppies are used to just walking in and telling me what they want and getting it. When they go to their new homes they are freaked out for a while, and can become shy and insecure. They don't know how to handle people not being able to understand it. Every dog in my home gets some time that belongs to them. The best time for me is in the early morning, when there are no phones ringing. I have twenty-two shepherds now, but some are rescues that I will place.

When I go to pounds I rescue shepherds. For example, I recently found this beautiful sable male shepherd, and I asked him, "What are you doing in here?" And he said, "My family is moving." He showed me pictures of the family packing and crying when the man put him in the car to bring him to the pound. He said, "They couldn't take me." I asked him if he had any training and he said, "Yes," and he showed me his being with children and loving them, and the man obedience-training him. So I said, "Are you afraid of anything?" and he said, "No." So I brought him home, and he was completely obedience-trained. I sent him to the police department in Florence, Oregon, where he is called DJ and is a first-class police dog.

I can also judge racehorses. I was recently back in Kentucky, and I saw a filly and said, "This horse is going to be a stakes winner, no question about it. She has a competitive spirit, she is brilliant and independent enough that even if she gets a bad trainer she is still going to go out there and win." And her owner said to me, "You know that filly you talked to in the womb before it was born, and you said she was going to be a stakes winner? Well, she's a three-year-old now, has already won four stakes races, and just sold for $450,000."

A horse can have the best bloodlines or conformation, but won't necessarily race well. Once a lady hired me to find out why her racehorse was always sick and injured. I asked the horse, "Do you hurt?" And he said, "No." So I said, "Why are you always limping?" And he said, "If I limp and act sore, I don't have to run, and I hate the track and I hate racing." They had spent $100,000 on this horse. When they heard what I had to say, they decided to turn him into a hunter jumper, but after they tried him a few times he kept refusing to jump. He told me, "I am not going over those sticks." But as he was standing there he was watching a dressage ring, and he said, "That is what I want to do." The people said, "But there's no money in dressage." Nevertheless, they decided to do it.

That horse was so good in just one month of training that he went to a show and took two firsts and a second. The judges came over and told the rider, "You are terrible, but the horse is so good we are giving it to him anyway. This horse is good enough for the Olympics."

I always have financial difficulties taking care of all my animals. Though the stress can really get to me, I would never live without them. They are my world, my business, and my life. I believe that one of the things God wants me to do is to make people aware that there is a different dimension to what he created. I believe, back in the Garden of Eden, Adam and Eve totally communicated with animals. Why else would Eve have talked with that serpent like it was an intelligent creature if it hadn't been a common thing to talk to an animal? And the animals were willing to go on the ark with Noah because there was still a bond with nature. But after the flood, the animals got a fear of man. I think God wants to restore that bond.

Something has to start balancing out this crazy world and somebody has to do it. That has been part of my mission. By communicating with animals you learn respect for life and treat all living things with more kindness and feeling. I don't like to call it psychic because that has the connotation of all the occult stuff — mostly dealing with spirits and predictions — and I'm not into any of that. This is just communication, nonverbal communication. I talk to the animals and they give me answers, the way other people talk to people. I am not dealing with a disembodied spirit, but with a mind that is alive and functioning.

*

These are some of Beatrice's "readings." The owners later corroborated the stories wherever it was possible for them to do so.

Your Doberman says he likes the ocean. He really liked a young guy he was very close to. He doesn't say that he was owned by a woman like you said. He doesn't even want to talk about the place you got him. He says he spent most of his life with a guy and won't even talk about the woman. He said, "The only thing I want to remember is that man and the happiness I had with him. We used to go camping. We did everything together." He said, "He left me with that lady, and he said he'd come back and get me and he never did." Did the woman have a boarding facility, 'cause I'm seeing kennels and things. I think his owner just never came back to pick him up. He really felt bad. That's a shame because they were good friends. Oh, that's heartbreaking. He was terribly depressed 'cause he kept looking for his master to come back all the time. He said he knows you really love him, but he has never let go of his previous owner. He hasn't really attached himself to you. He's coming around, but it's going to take him a little time. I wonder whatever happened to that guy? I hate to tell you this because I know you're very fond of the animal, but he really still grieves for that man. Maybe you can track him down through a forwarding address on his papers and at least find out what happened to him.

This Dalmation said she was born in a yard, not a kennel. She said the people kept her outside and she would look in through a glass door. She said, "They would let me in the kitchen but didn't want me in the rest of the house. Then they were packing and moving and I had to go." They gave her to somebody with a wood fence. She said, "All I did was cry. I hated it. I cried and cried and stayed inside this high wooden fence. I was lonely." She tore up a cushion on the back porch. "I chewed on anything I could find, including the gate, the fence, and the door." She got out and started running and was put in the pound. What are you feeding her? Those moist package foods are one-third sugar and salt. But she said, "They think I'm too fat, but I'm not too fat." She really loves her food and doesn't want you to cut her down!

This cat sits on the counter most of the time, doesn't she? I see her sitting up on something that looks like a counter. It doesn't look like a kitchen counter. It's brown and about this wide . . . the stereo cabinet. She said, "That's my favorite spot."

This parakeet, did he fly into your place? Because when I said, "Where did you come from?" he said, "I flew out the window and then they found me in the yard. They didn't buy me; the other people did." He was already tame, too, because he said he was used to people being around him. Is the thing you're trying to find out connected with the cat? I can feel fear, a lot of fear, connected with the cat. Did you have a built-in cage or something? It's a big thing, almost like something one would create rather than buy in a store. Was there a green parakeet in there? The cat ate your green bird, didn't it? 'Cause I said, "Where did the green bird go?" And he said, "There was a lot of racket, and the next thing I knew there were feathers all over." If you hadn't come around when you did, he'd have gotten this one too. The other one talked, didn't it? 'Cause he said, "That one really talked." They were good friends. He says there was a lot of flying around. Is there a big window in the room, either a big window or a sliding glass door? 'Cause he said, "The green bird flew into the window and fell to the floor, and the cat got him. I got back in the cage, and the other one was flying and hit the glass and fell, and that was when the cat got him."

The horse remembers jumping and being happy at it. He said, "The person who was jumping me didn't come back." He was a man, wasn't he? Well, he didn't know what happened, and he was wondering if it was something he did. It left him a little unsure of himself. Now he says he's not exactly sure of what he's supposed to be doing. I can feel through him a hesitation on your part when you're approaching a jump — like all of a sudden you don't feel confident. Then he starts getting confused. He said his owner didn't do that. He knew exactly what he wanted. You see, there are two types of horses. You've got the kind that gets confidence from the rider and needs to be told. The other kind knows what he's doing and is gonna do it anyway, and the rider doesn't upset him. I think as you gain more self-confidence and more assurance, the more the two of you will perform better. He enjoys the jumping; he just needs you to help him. He's satisfied with things, but he did certainly love his former owner.

Your horse says she wants a baby and that she really loved a baby another horse had. Did she come off a big thoroughbred farm somewhere? 'Cause I'm seeing a big farm, and she said, "I was really happy there." She says you give her an awful lot of love and attention, and that feels good. She says she tries to comfort you sometimes, and that you were going through a spell and were really upset. She said, "I tried to tell her I loved her and I was taking care of her." She is a very sensitive animal and very smart.

Animal Child

All the animal people I have talked to have told me that they had special feelings toward animals from their earliest memory. Talking to an animal child was a way of tracing the river to its source.

Shannon Lasdon, 9½ years old, lives with her mother, Marsha, her veterinarian father, Foster, and her younger sister, Alida, in a luxurious three-bedroom house in the San Fernando Valley. Foster is an animal person, and most of the animals living inside and outside the house are his or Shannon's. In the living room, on their perches, sit Paco, a double yellow-headed Amazon parrot, and his companion, Bronco, a Nandae conure. Brew, a mixed-breed terrier, has free run of house and yard. Kittens play in a carpeted multicompartment cat habitat that reaches almost to the ceiling. Stretch the tarantula lives in the den, which also contains aquariums full of freshwater fish, including some rare ones such as the black ghost.

The large backyard is lushly planted. A swimming pool occupies one side. Goldfish swim in water-filled half-barrels. An aviary with breeding pairs of parakeets is pressed against the back of the house. Wire cages with rabbits inside are on the patio and on the far side of the yard.

A spacious enclosure, beautifully landscaped, runs along the length of the large swimming pool. In it live quails, chickens, doves, pheasants, mandarin ducks, and homing pigeons. Seven individual cages, each over 6 feet tall with natural greenery, house rosellas and button and silver dove quail. A squirrel lives in her own cage, and three pairs of fancy pigeons and their babies are separately housed. Before problems arose with licensing, Misty the cougar used to live in a huge cage along the side of the house.

SHANNON LASDON: I was born in Los Angeles. I'm nine and a half. I'm in fourth grade. My sister is eight. All she likes is cats. Before Mommy married my dad, he had a little red fox and a raccoon. My grandmother told me all this.

My mom is not really interested in animals. Like when we go to bird club meetings, it's just me and my dad. My sister just likes the refreshments, but I listen to the speeches and look at the birds.

The first animal I had was a parakeet. It was blue and white and had black dots and a purple beak. My dad picked him out. He takes care of tigers, lions, snakes, and turtles, but mostly he takes care of common animals like dogs, cats, and little pets like rabbits, guinea pigs, and hamsters. He loves animals, especially birds.

I have two favorite kinds of animals, and I can't really make up my mind   199 which I like best — cats or rabbits. Both are soft and cuddly and cute. When I was younger, most of my dreams were about rabbits talking to squirrels. Then they were my favorite animals because I didn't know they would bite. I pictured a squirrel that would talk and wouldn't bite. The squirrel would be sharing some peanuts and the rabbit would be picking some carrots, and they would be talking about having a dinner party.

I liked Misty, the mountain lion we once had, but sometimes I was a little scared of her. When she lived at our house, almost every day I would go outside and say hello to her and rub my hand across her cage to pet her. When I was born my dad had her, and she was just a little baby. My dad has pictures of me when I was a baby playing with Misty in the living room on a big white carpet. I can remember a little bit. One time I was playing with her and she started to stretch out, so I licked her paw. Kids in my class make up stories like "I used to have a tiger," but they really didn't. But I really had a mountain lion.

I always wish I could have a zebra or a giraffe. I'd like to look at the giraffe and pet it. I want a zebra because it is kind of like a horse and pretty. I would also like a swan, a nice white swan that has wings that go up.

I am allowed to go in the cages and get chicken eggs and pigeon eggs. One day I saw one pigeon sitting on chicken eggs and pheasant eggs. When the pigeon got up, I just collected the chicken eggs. The chicken eggs we put in the refrigerator, and we have them for breakfast. The other eggs, like the pheasant eggs, we put in the incubator to hatch. My favorite part about incubating eggs is when they hatch. Little tiny black birdies run around and squeal. While we are playing in the living room we can suddenly hear little chirps, and we know they've hatched.

We had a Newfoundland dog. He was so big and black, everyone called him a bear. He was really old and was really sick. Two years ago he died during surgery. His name was Becket and I loved him. He loved to play and he slobbered a lot. My sister tried to ride him, and she sat on his back and fell off. When I took him for a walk, people would come up and pet him and he would sit down and just kind of stare at them with a smile. Now we have Brew. He is a Benji dog and my dad got him.

I have seen animals that died. One time a birdie was stuck in the egg, and my daddy tried to take it out. He got it out and he put it under a hot light. It didn't make it. I got really upset. One day I went outside and the white rabbit was lying dead on the ground. It was a really hot day, and Brew had chased it around. It died of fear and heat.

A lot of my responsibility is to take care of the bunnies. I take care of the squirrel and take care of the fish. I name most of the animals. The mother cat has about fifteen different names. I named some of the chickens and all the rabbits. I think animals have a kind of reaction like people do. If someone says something exciting, a person will act excited, and if a rabbit sees something it likes, it will get real excited too.

I would like to be an animal trainer when I grow up. I was thinking of being a veterinarian, but there are some things that a veterinarian does that I don't think I would like — like when an animal dies or when you have to hurt it to make it better, like surgery. It gives me the chills.

Right now, because I have so many animals, animals are really a part of my life. If someone came and took them away, it would be like someone taking part of my life away. If I didn't have animals, every time I saw an animal I'd beg for it and just be kind of sad.

Am I an Animal Person?

Saying goodbye to an animal person at the end of the interview always left me sad. Again and again I experienced the need to brace myself the way one would on moving from a warm hearth to a wintery chill. The animate world as seen by animal people was so sentient and vibrant that the world as perceived by other people was quite bleak by comparison. I also missed their animals, creatures never burdened by being identified as pets or property. My solace came from recognizing that in meeting these people I had become more like them. My loss was never total.

Early in my travels, I felt cravings to own many of the animals with which the animal people lived. I had fantasies of miniature horses in the front yard, my swimming pool converted to a koi pond, terrariums filled with spotted salamanders, and a breeding colony of Giant Madagascar Hissing cockroaches. I did not yield to temptation, however, because I had learned from the animal people that proper care of these creatures requires an intensity of devotion and time that I simply do not have. And in my decision not to get more animals I think I transcended the petkeeper part of my mentality and became even more of an animal person. For a petkeeper the urge is to possess, no matter what the consequence to the animal. But to an animal person, an animal is so real in its experience of hunger, fear, or loneliness for its own kind that it is not morally or emotionally possible to live with one unless the conditions are ideal.

My lesson can be summed up by one word: intimacy. I became convinced that all living things can experience their interconnectedness.

One summer day I was sitting in my dining room, and a bottle green fly buzzed by and made a vain attempt to penetrate the glass of the large picture window. Without giving it much thought, I said to the fly, "If you would like to leave, I'll take you out." To be helpful, I extended my index finger but made no other change in position. The fly circled me and landed on my finger. I continued, "You will have to trust me, because I will have to stand up and walk though three rooms in order to let you out through a glass door." And the fly held his place, and when the door slid open, out it flew, leaving me with tears in my eyes. Could the fly understand?

Then I remembered an incident from many years before, while I was in medical school. There was a commotion on the roof of one of the grand, four-story limestone buildings in the quadrangle. I was told that a German shepherd used for dog lab had escaped from his cage and run up the stairs to the roof. "Dog lab" is the euphemism for the vivisection that is done on dogs by medical students to gain training in surgical techniques. I later learned that the dog ran to the ledge, looked back at his pursuers, and then leaped to his death. It seemed quite plausible that the dog had committed suicide.

But many people are not animal people. On one of my transcontinental flights, I met a man returning to his native Africa. When we spoke about animals, he stated that he would like to tell me his point of view. His story helped me to understand that being an animal person is one of the luxuries of not having to struggle for survival. As he said, "Maybe more people can be animal people if they have enough to eat."

TRAVELER: I was born in Ghana in 1943 and lived there almost all my life. I have my point of view about the animals being killed in Africa.

I put it this way — the jungles where the animals live are getting smaller 'cause the villages are getting bigger. We cannot sit around fearing that the animals will come and destroy us, so people say we got to kill them before they kill us — elephants, lions.

I remember years ago, the elephants are so many we couldn't even get close to them and the villages were small, maybe one village would have fifteen to twenty people; but now the population is getting bigger. Before we don't have the guns to protect ourselves.

Six years ago there was a small village in the north part of Ghana. The elephants come over there and demolish the whole village, but nobody die — the people run away. The elephants coming in and there is nothing the people can do to stop them. They don't have guns or anything. They left the village and the houses empty, and the elephants destroy the whole houses. They went to the nearest village where they had guns to borrow some to come back. The elephants no come back, but the people were ready for them to come.

We do have guns twenty years ago, but the villagers are not rich enough to buy a gun. Now they afford to buy guns. Now they kill the animals to sell parts. The elephants, they sell the tusks. But from my point of view, the main problem is the tourists, 'cause they come in there and they want to buy some animal something, and the villagers go into the jungle, they kill the animals to sell something.

There are not as many elephants as there used to be because they kill them and sell their ivories and stuff. We don't have that many lions. Years ago, you can walk 4 or 5 miles in the jungle and maybe you could meet up with four or five lions, but not anymore. You have to be two or three days and you can't find any. We used to go out there and try to hunt not with guns — this was a long time ago — we used bows and arrows. It is difficult because when you shoot this animal they don't die at the moment you shot them, so you have to follow them and the only way to trace them is by the blood. We have the old hunter there, and he teaches us how to track an animal.

If you live in the jungle with your uncle or your father, they have to teach you. Sometimes it is the only way to survive, to get food to eat. No food gets into the jungle where the villages are. We have to grow most of our own food or we have to hunt. There are no cattle. One tribe in Ghana, they eat elephant meat. They can't kill the bigger ones, so they have to kill the babies. The babies are easy to kill.

In Africa, people don't believe in animals like people in the Western world. We been living with animals all our lives and we were afraid they could kill us. We don't keep them as pets. We don't have big zoos. Over there, many people don't even think about saving the animals. They say, "The less animal the better, because nothing come to try to hurt us or kill us." I don't see no reason to be saving wild animals that can kill you. If you don't live with them and you only go to a zoo to take a look at them, you would like to save them. But if you live where they can harm you, then you don't want to save them. We can't build a big fence all over the jungle and feed them.

It would be difficult to go to any house in Africa and see pets like you do here in America. Many people have no love for animals. Even the dogs are wild. So it is strange for me to go to a house in America and see a dog sitting in the living room or riding in cars. Thirty miles from my tribe people eat cats. Some tribes still eat dogs. Most people in America don't realize that half of the animals you keep as pets, we eat them.

This guy I know came from Africa and he came and visit a guy. And the guy was going to work, so he left the apartment for him. So he went walking around the street and he saw the squirrels. And he went and killed one, went back to the house, put it on the stove, and tried to burn the skin off. So there was a lot of smoke there, so the smoke detector went off and the fire department came. And when they came he was busy skinning the squirrel. "What are you doing?" "I am doing? I am skinning this to eat it." They say, "We don't eat them in America." And he says, "It is unbelievable that you see these things walking on the street here. The hunters go to the jungle all day long just to hunt, something like that, and here they keep it as a pet."

203

I visited Michael Crichton because he wrote *CONGO,* a novel that features subtle and tender descriptions of sign language communication between a young gorilla and her human teacher and friend. I was not sure whether or not Michael Crichton was an animal person.

By the time of the interview, I had learned from my many conversations with animal people that their telltale common denominator was an early childhood memory of a yearning to be close to some other species. But Michael said to me, "My background was quite lacking in any contact with animals. I never had a pet when I was growing up, and I was never very anxious about it. I grew up in a suburb outside New York. There were no wild animals around. I was raised in a man-made environment surrounded by people."

He said he liked his golden retriever, Foxy, but had no particularly strong attraction to nonhumans. So Michael Crichton was not an animal person. Nevertheless, his insights added to my belief that humans have a fascination for the nonhuman. He expressed an animal-person-like affection for his Olivetti word processor and his IBM computer. It occurred to me that in some hypothetical future time, long after other creatures are gone, people could still bond to different "species," albeit sentient machines rather than flesh-and-blood types, for solace and an opportunity to transcend their boundaries of humanness. He told me:

MICHAEL CRICHTON: Recently I had to go for a casting call for a film version of *CONGO.* Instead of actresses and actors, apes were brought into a little trailer. To see apes in close quarters, not in a zoo setting or in a wild setting like Africa, but in a room with walls and doors and telephones ringing, was disturbing. They are hairy, smelly; their hands are oddly shaped; their legs are too short; and their trunks and arms are too long. But there is something about them that is too close to us. They provoke a kind of discomfort that we never feel around a dog. Dogs are great; dogs are animals. No question; dogs are different from us.

I don't think people like to be reminded of their animal nature. I can remember when I began to dissect a cadaver. Of all the things I thought would really disturb me being around a dead person — getting flesh under my fingernails, the smell, not feeling like eating lunch after the work — all those things that I feared turned out to be only moderately unpleasant. But the shock was to open up bodies and see that they contained elements you would see in a butcher shop. It was coils of intestines and organs. There wasn't anything in there that I had never seen from animal work. I had a sort of real desire that there be something unique in the human body, and in that sense dissection was very distressing and constantly very disappointing.

The implication, of course, is there is nothing special about us. We are just animals, and whatever separates us from those chimpanzees that were brought into the trailer looking for a job in *CONGO* is not enough to be really satisfying. I can understand people's vested interest in specialness, but I am not as jealous as many. Some years ago, when I heard that the chimpanzee Washoe had begun to use sign language, I thought, "Absolutely." That seemed correct to me. I had thought that there ought to be a spectrum of consciousness, and I was relieved to hear that there was.

One day at the beach, a little kid came over and asked me whether I was allowed to go in the deep end. He thought the ocean was a pool. The second thing he said was, "When do the waves go to sleep?" He had trouble with the idea that it was ceaseless, because everything else stopped and went to sleep as far as he was concerned. I remember looking at this kid and thinking, "That is something an adult would never think of." In their perceptions, children will always bring you up sharply and pull you out of your adult world, make you honest in some way. Animals can do that also. Just because they are different.

Earth still flourishes with myriad species. For the time being, the option still remains for animal people and non-animal people to experience the rich, primordial heritage of the planet. Wilderness and jungles still impinge on our cities and our conscious-ness. We humans and the surviving animals, whose genealogies also span millions of years, are poised in a delicate balance. Each day echoes, "Who is to die, and who is to live?" An evolutionary drama is unfolding. In their own ways the animal people see it happening, and they care. For if animals become extinct, so do they.